QUANTITATIVE THERMOCHRONOLOGY

Thermochronology, the study of the thermal history of rocks, provides an important record of the vertical motions of bodies of rock over geological timescales, enabling us to quantify the nature and timing of tectonic processes. Isotopic age data constrain the ages of rocks and minerals, but in many cases they are interpreted without a proper understanding of the relationship between the age measured and the physical processes within the Earth.

Quantitative Thermochronology is a robust review of the fundamental nature of isotopic ages, and presents a range of numerical modelling techniques to allow the full physical implications of these data to be explored. The authors provide analytical, semi-analytical and numerical solutions to the heat-transfer equation in a range of tectonic settings and under varying boundary conditions. The second part of the book illustrates their modelling approach, which is built around a large number of case studies. Various thermochronological techniques are also described in order to help the non-specialist understand the benefits of each method.

Computer programs that provide a means of solving the heat-transport equation in the deforming Earth and allow the prediction of rock ages for comparison with geological and geochronological data are available on an accompanying website (www.cambridge.org/9781107407152). Several short tutorials with hints and solutions are also provided.

JEAN BRAUN is Professor of Geosciences at the Université de Rennes 1, Adjunct Professor at the Research School of Earth Sciences at the Australian National University and Fellow of the Canadian Institute for Advanced Research. His research interests include the quantitative study of continental tectonics, including lithospheric and crustal deformation; landform evolution and its feedback on tectonics; glacial erosion; soil transport; thermo-mechanical interactions in the crust; thermochronology; rock rheology; fluid flow in fractured rocks; and the development of numerical methods for the solution of Earth problems.

PETER VAN DER BEEK is Lecturer (Maître de Conférences) in Earth Sciences at the Observatoire des Sciences de l'Univers de Grenoble, Université Joseph Fourier, Grenoble, France. His research focusses on the interaction between tectonics and surface processes; numerical modelling of erosional processes; long-term landscape development; and quantifying exhumation and erosion histories using fission-track thermochronology and cosmogenic nuclides.

GEOFFREY BATT is Senior Lecturer and ExxonMobil Teaching Fellow in the Geology Department, Royal Holloway, University of London. He is both active in thermochronological research and a committed educator and member of the British Higher Education Academy. He has been awarded prizes for his teaching, in particular relating to enhancing learning and knowledge exchange through an online learning environment. His research focusses on the tectonic evolution of plate-boundary regions through competing constructional and exhumational processes.

QUANTITATIVE THERMOCHRONOLOGY

Numerical Methods for the Interpretation of Thermochronological Data

JEAN BRAUN
The Australian National University, Canberra, Australia
Now at Université de Rennes 1, Rennes, France

PETER VAN DER BEEK
Université Joseph Fourier, Grenoble, France

GEOFFREY BATT
Royal Holloway, University of London, United Kingdom

CAMBRIDGE
UNIVERSITY PRESS

University Printing House, Cambridge CB2 8BS, United Kingdom

Cambridge University Press is part of the University of Cambridge.

It furthers the University's mission by disseminating knowledge in the pursuit of education, learning and research at the highest international levels of excellence.

www.cambridge.org
Information on this title: www.cambridge.org/9780521830577

© J. Braun, P. van der Beek and G. Batt 2006

This publication is in copyright. Subject to statutory exception and to the provisions of relevant collective licensing agreements, no reproduction of any part may take place without the written permission of Cambridge University Press.

First published 2006
First paperback edition 2012

A catalogue record for this publication is available from the British Library

ISBN 978-0-521-83057-7 Hardback

Cambridge University Press has no responsibility for the persistence or accuracy of URLs for external or third-party internet websites referred to in this publication, and does not guarantee that any content on such websites is, or will remain, accurate or appropriate.

Contents

Preface	*page* ix

1	**Introduction**		**1**
	1.1	Thermal history: the accumulation of thermochronological age	3
	1.2	Cooling, denudation and uplift paths	7
	1.3	Thermochronology in practice	13
2	**Basics of thermochronology: from t–T paths to ages**		**19**
	2.1	The isotopic age equation	19
	2.2	Solid-state diffusion – the basic equation	20
	2.3	Absolute closure-temperature approximation	23
	2.4	Dodson's method	24
	2.5	Numerical solution	27
	2.6	Determining the diffusion parameters	30
3	**Thermochronological systems**		**33**
	3.1	Ar dating methods	33
	3.2	(U–Th)/He thermochronology	42
	3.3	Fission-track thermochronology	48
4	**The general heat-transport equation**		**60**
	4.1	Heat transport within the Earth	60
	4.2	Conservation of energy	61
	4.3	Conduction	63
	4.4	Advection	64
	4.5	Production	65
	4.6	The general heat-transport equation	66
	4.7	Boundary conditions	66
	4.8	Purely conductive heat transport	68

5	**Thermal effects of exhumation**	**76**
	5.1 Steady-state solution	76
	5.2 Thermal effects of exhumation: transient solution	81
	5.3 Thermal effects of exhumation: the general transient problem	83
6	**Steady-state two-dimensional heat transport**	**105**
	6.1 The effect of surface topography	105
	6.2 The age–elevation relationship – steady state	110
	6.3 Relief change	113
7	**General transient solution – the three-dimensional problem**	**115**
	7.1 Pecube	115
	7.2 Time-varying surface topography	116
	7.3 Surface relief in the Sierra Nevada	118
8	**Inverse methods**	**122**
	8.1 Spectral analysis	122
	8.2 An example based on synthetic ages	124
	8.3 Application of the spectral method to the Sierra Nevada	127
	8.4 Sampling strategy	129
	8.5 Systematic searches	130
9	**Detrital thermochronology**	**131**
	9.1 The basic approach	131
	9.2 Deconvolution of detrital age distributions	136
	9.3 Estimating denudation rates from detrital ages	140
	9.4 Estimating relief from detrital ages	144
	9.5 Interpreting partially reset detrital samples	148
10	**Lateral advection of material**	**151**
	10.1 Lateral variability in tectonically active regions	151
	10.2 Exhumation and denudation in multi-dimensional space	152
	10.3 Consequences of lateral motion for thermochronology	153
	10.4 Scaling of lateral significance with closure temperature	154
	10.5 Evaluation of the significance of lateral variation	155
11	**Isostatic response to denudation**	**164**
	11.1 Local isostasy	164
	11.2 Flexural isostasy	166
	11.3 Periodic loading	167
	11.4 Isostatic response to relief reduction	168

11.5	Effects on age distribution	168
11.6	Effects on age–elevation distributions	170
11.7	Application to the Dabie Shan	171

12 The evolution of passive-margin escarpments — **177**
12.1	Introduction	177
12.2	Early conceptual models: erosion cycles	179
12.3	Thermochronological data from passive margins	180
12.4	Models of landscape development at passive margins	182
12.5	Combining thermochronometers and modelling	186

13 Thermochronology in active tectonic settings — **192**
13.1	A simple model for continental collision	192
13.2	Heat advection in mountain belts	196
13.3	The Alpine Fault, South Island, New Zealand	199
13.4	Application of the Neighbourhood Algorithm to Southern Alps data	202

Appendix 1	*Forward models of fission-track annealing*	207
Appendix 2	*Fortran routines provided with this textbook*	210
Appendix 3	*One-dimensional conductive equilibrium with heat production*	211
Appendix 4	*One-dimensional conductive equilibrium with anomalous conductivity*	214
Appendix 5	*One-dimensional transient conductive heat transport*	216
Appendix 6	*Volume integrals in spherical coordinates*	220
Appendix 7	*The complementary error function*	222
Appendix 8	*Pecube user guide*	224
Appendix 9	*Tutorial solutions*	228
References		237
Index		255

Preface

The Earth's surface is continuously reshaped by the interaction of tectonic and surface processes. Where the tectonic forces acting on the lithosphere lead to downward vertical motions or subsidence, the resulting depressions are usually filled with sediments that contain a record of these vertical motions. In actively uplifting regions, however, as well as in passive but formerly uplifted regions of relatively high topography, the surface process response will be mostly erosional and no direct record of past vertical motions exists. In such systems, thermochronology – the study of the thermal history of rocks – provides practically the only record that can be obtained in terms of vertical motions on geological timescales. However, this record is highly non-linear and depends on many parameters that need to be understood in order to interpret thermochronological data meaningfully. In particular, one needs to understand: (1) the relationship between the thermal history of a rock and the accumulation of thermochronological 'age', as well as the influences of various physical and chemical parameters on this relationship; and (2) the relationship between advection of rocks towards the surface by the combined effects of tectonics and surface processes, and the thermal structure of the rocks (i.e., the links between the thermal and structural reference frames).

Several outstanding and fundamental problems in the Earth Sciences will rely partly or entirely on the meaningful interpretation of thermochronological datasets for their resolution. To name but one, the debate about the possible interactions between Late Cainozoic climate change and the uplift of some of the Earth's largest mountain belts has been going on for the last 15 years. Several independent datasets show that sedimentation rates in many of the world's sedimentary basins have doubled or tripled over the last 2–5 Myr (e.g., Métivier et al., 1999; Zhang et al., 2001; Kuhlemann et al., 2002). This increase is traditionally ascribed to tectonic surface uplift of major mountain belts, which would have led to drawdown of atmospheric CO_2 through increased rates of rock weathering, and ensuing climatic cooling (e.g., Raymo and Rudiman, 1992). However, as Molnar and England (1990) and Zhang et al. (2001) point out, the lack of independent evidence for surface uplift of mountain ranges during this timespan and the widespread nature of the increase in sediment flux suggest the driving mechanism for this increase to be climatic; thus, the uplift of mountain peaks may be an isostatic response to, rather than a trigger of, increased denudation rates. Such a mechanism requires denudation rates to be highly spatially variable: valley bottoms must erode much more rapidly than mountain peaks (i.e., relief must increase) in order for isostatic rebound to be effective in uplifting the peaks (e.g, Montgomery, 1994; Small and Anderson, 1998). The resolution of this debate will thus depend in part on our ability to constrain temporal and spatial variations in exhumation rates within eroding mountain belts effectively.

A related question has arisen out of the realisation that tectonics and erosion are not independently operating processes but must be strongly coupled (e.g., Beaumont *et al.*, 1992; Zeitler *et al.*, 2001). In effect, while tectonics controls erosion rates through the creation of surface relief, erosion in turn also affects tectonic patterns through its role in displacing mass at the surface, thereby influencing the thermal (and hence rheological) stress and potential-energy fields of actively eroding regions. The question is that of whether this coupled system is fundamentally controlled by an internal (tectonics) or external (climatic) driving force. Authors who have attempted to address this question through the comparison of spatial patterns of thermochronological ages with present-day climate (precipitation) data have come to conflicting conclusions (e.g., Burbank *et al.*, 2003; Reiners *et al.*, 2003a; Thiede *et al.*, 2004; Wobus *et al.*, 2003), and its resolution will depend in part on a better comprehension of the significance of spatial variations in thermochronological ages across orogenic systems.

A third and final example may be drawn from the debate about the relative timescales of tectonic versus erosional processes and the significance of the concept of topographic steady state. From a theoretical viewpoint, it is relatively simple to demonstrate that active tectonic systems that are subject to continuous uplift at a constant rate should tend towards flux steady state, where the tectonic influx of material is balanced by the erosional outflux (e.g., Beaumont *et al.*, 1999; Willett *et al.*, 2001). However, it is not clear whether such systems also reach denudational and topographic steady state (where denudation rates and surface topography, respectively, are constant in time). Thermochronological data can be used to test thermal and denudational steady state in mountain belts (Willett and Brandon, 2002), but attempts to demonstrate the existence of denudational steady state in natural settings from such data (e.g., Bernet *et al.*, 2001) have been criticised under the argument that they were inconsistent with other thermochronological datasets (e.g., Carrapa *et al.*, 2003; Cederbom *et al.*, 2004). Again, resolution of this question will depend on our understanding of the significance of spatial and temporal patterns in thermochronological data.

Technological advance has seen the accuracy and precision of thermochronological techniques rise steadily over the past 15 years. In parallel with this, understanding of the fundamental behaviour of isotopic chronometers on geological timescales has progressed to the point where the thermal sensitivity of key systems can be constrained on a routine basis. These advances have brought progressively greater benefit to quantitative interpretation as a way of adding value to what remain costly analytical techniques.

The purpose of this book is to provide tools to help the Earth scientist undertake such interrogation of thermochronological data in a rigorous and quantitative manner, and, in particular, to extract the information such data contain on the tectonic (i.e., internal deformation) and geomorphic (i.e., surface evolution) history of a given region.

Although working from generally applicable principles, the techniques we develop and describe here are targeted most directly at the interpretation of thermochronological data in environments where (a) tectonic uplift and erosion combine to bring rocks to the Earth's surface, where they are collected and analysed by the geologist and (b) the amplitude of surface relief is important. We focus in particular on the effect of rock advection and finite-amplitude surface relief on the temperature structure in the Earth's crust and consider a range of situations from the simplest, where tectonic advection is so slow that the heat is mainly transported by conduction, to the most complex, where rapid vertical advection of heat through a convoluted, cold upper surface produces a complex three-dimensional thermal structure that must be considered. We also consider a range of tectonic settings, ranging from situations where rocks are actively uplifted by ongoing tectonic processes and erosion counteracts the resulting uplift of the surface to situations where, in the absence of any tectonic forcing, rock uplift is solely driven by the isostatic rebound caused by erosional unloading.

The book is built around a progressive increase in complexity both of the tectonic scenarios considered and of the interpretative tools consequently required to interpret thermochronological data. Along the way, a set of Fortran programs developed by the authors is progressively introduced, together with an accompanying series of tutorials designed to assist the reader in understanding how to use these tools to interpret the increasingly complex scenarios. The Fortran programs can be downloaded from the web page; their application will allow the reader to solve the tutorial problems. More importantly, however, the programs are designed to be applied to the interpretation of real thermochronological data in different settings. The Fortran program **Pecube**, which is presented towards the end of the book, is a very general solver of the heat-transport equation that includes conduction, production and advection of heat, as well as the option of varying the geometry of the surface to represent geomorphic processes. It has been designed for ease of use by any geologist interested in the interpretation of thermochronological datasets. In this context, the book can be regarded as a long, and, we hope, not too painful, introduction to this powerful modelling engine, providing a numerical 'toolkit' for a geoscientist looking to interpret such data.

In designing and writing this book, we are aiming at an audience of upper-level undergraduate students and graduate students who are commencing their studies. We have included a set of slides (in PDF format) (see the website) for use during lecturing and tutorials, if one wishes to use this book as a teaching tool. We assume that our audience is acquainted with the basic principles of geochronology, as well as the basics of geodynamics and heat transport in the Earth's crust. Although we review these basics in early chapters, readers who are unfamiliar with this material may wish to consult textbooks developing these issues in more detail, such as Faure (1986), Turcotte and Schubert (1982) and Fowler (2005). Developments in the field of thermochronology and, more widely, the study of the interaction between tectonics and surface processes are rapid; while we have aimed this book to be up to date as it appears, it is inevitable that we fail to include the most recent studies, however important their implications may be.

The idea for this book was triggered by a short course given by Jean Braun during his appointment as visiting professor at the Université Joseph Fourier in 2002. Most authors who have written a textbook will admit that it has taken them about two to three times as long as they had initially planned. This book is no exception to the rule and we thank the staff at Cambridge University Press for their patience and understanding. Writing of the book was facilitated by mutual visits of the authors; the Research School of Earth Sciences at the Australian National University and the Observatoire des Sciences de l'Univers de Grenoble at the Université Joseph Fourier provided logistical support during these stays. Our thinking on the interpretation of thermochronological datasets was shaped in part through thought-provoking discussions and exchanges with colleagues, collaborators and (former) graduate students. We would like to acknowledge in particular Matthias Bernet, Mark Brandon, Roderick Brown, Jim Dunlap, Kerry Gallagher, Frédéric Herman, Barry Kohn, Erika Labrin, Ian McDougall, Peter Reiners, Xavier Robert, Malcolm Sambridge and Michael Summerfield. Many of the ideas presented in this book were developed during research projects funded by the Australian Research Council (ARC), the Institute of Advanced Studies of the Australian National University, the Institut National des Sciences de l'Univers (INSU) of the Centre National de la Recherche Scientifique (CNRS) in France and the National Environmental Research Council (NERC) in Britain. Finally but definitely not least, we would like to thank our partners Myriam, Gianna and Victoria as well as our families for their support and understanding as we became more and more submerged in this undertaking.

1

Introduction

Thermochronology is a technique that permits the extraction of information about the thermal history of rocks. It is based on the interplay between the accumulation of a daughter product produced through a nuclear decay reaction in the rock (whether this daughter product be an isotope or some sort of structural damage to the mineral lattice) and the removal of that daughter product by thermally activated diffusion. Because temperature increases with depth in the Earth's lithosphere, this temperature information can be translated into structural information – thermochronological data thus contain a record of the depth below the surface at which rocks resided at a given time. For eroding basement terrains, where rocks have been brought to the surface from depths of several to several tens of kilometres, thermochronology is the only technique that will provide such information and permit one to constrain the timing of rock exhumation towards the surface quantitatively.

However, the relationships among vertical motions, temperature history and the accumulation of daughter product (the present-day abundance of which will provide a thermochronological age) are highly non-linear and depend on many parameters. To the Earth scientist seeking to constrain geological history, this presents a dichotomy. On the one hand, such thermochronological datasets have the potential to offer insight into how parameters that influence the thermal history of a geological sample vary over time. On the other hand, however, to extract information on any one aspect of the system requires an explicit understanding of the interplay of all of the variables. In particular, one will need to worry about how the accumulation of daughter products is affected by varying temperatures (and thus understand solid-state diffusion in minerals), but also about how tectonics and surface processes affect the thermal structure of the lithosphere, in order to comprehend the links between the thermal and structural frames of reference (Figure 1.1). As an example, consider the thermochronological dataset that is plotted in Figure 1.2 as the age of a sample versus the elevation at which it was

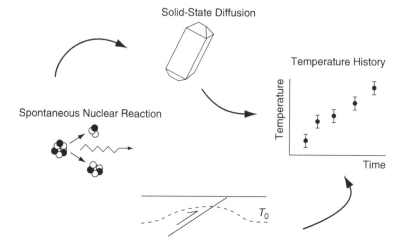

Fig. 1.1. Interpretative steps in relating thermochronological ages to structural history. The measured age is controlled by the accumulation history of daughter products in the host mineral, which depends on the thermal history. The thermal history is controlled on the one hand by the exhumation history of the rock, and on the other hand by the thermal structure of the crust. These two factors are interdependent: owing to the advective component of heat transport, the thermal structure depends on the exhumation history, which itself may be guided by differences in thermal (and therefore rheological) structure.

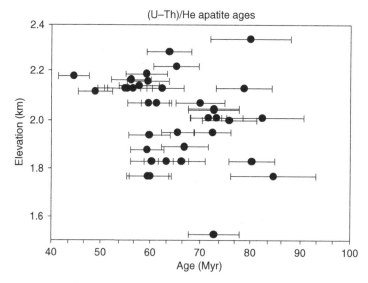

Fig. 1.2. Apatite (U–Th)/He ages from the Sierra Nevada, California, plotted versus the elevation at which the samples were collected. Data are from House *et al.* (2001).

collected. The ages are for the apatite (U–Th)/He system, which is sensitive to temperatures of ∼40–70 °C, which are typically encountered at depths of 1–3 km within the crust. Plotted in this manner, the data are slightly bewildering – they appear very scattered and no obvious relationships appear. There can be many reasons for this: the individual samples may be sensitive to varying temperatures and therefore not record passage of a single isotherm in the crust; exhumation rates and/or thermal structure may vary laterally; the surface relief may influence thermal structure and this influence may vary with time. The challenge is to find out what reason or combination of reasons fundamentally controls the observed spatial distribution of ages and thereby constrain the geological history of the region studied.

The purpose of this book is to provide tools to help the Earth scientist undertake such interrogation of thermochronological data in a rigorous and quantitative manner, and, in particular, to extract the information it contains on the tectonic (i.e., internal deformation) and geomorphic (i.e., surface evolution) history of a given region.

Alternatively, this textbook may be regarded as an exercise in quantitative or numerical modelling. The problems addressed are relatively well posed and we can therefore make use of a range of mathematical methods, ranging between two end-members:

- deriving the exact analytical solution to a simplified form of a general equation, which must then be regarded as an approximation of the real Earth problem; and
- deriving a numerical solution to the general form of an equation, which might be a better representation of the Earth problem but generally provides less insight into the behaviour of the physical system.

Both approaches are developed here, where possible and practical.

1.1 Thermal history: the accumulation of thermochronological age

Rather than reflecting a specific geological event, the isotopic ages of rocks and minerals are fundamentally determined by their integrated thermal history. Every isotopic system will behave as an open system at high temperatures, at which the daughter product is removed by diffusion more rapidly than it is produced by nuclear decay, and as a closed system at low temperatures, at which removal is so slow that all daughter product is retained within the host mineral over geological timescales (Figure 1.3).

The switch from open- to closed-system behaviour is not instantaneous, but rather takes place over a discrete temperature interval known as the *partial-retention zone*. Somewhere within this temperature range lies the *closure temperature*, formally defined by Dodson (1973) as the 'temperature of a

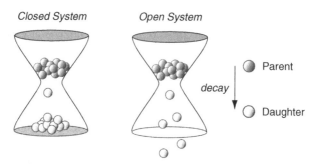

Fig. 1.3. Hourglass analogies to open- and closed-system behaviours. In a closed system (left), the daughter product accumulated by nuclear decay of a parent isotope is completely retained within the host mineral; an isotopic age builds up linearly with time. In an open system, the daughter product is removed effectively immediately upon its production; no isotopic age is built up within the system.

thermochronological system at the time corresponding to its apparent age'. The closure temperature provides the simplest conceptual entity to which a thermochronological age can be related; from its definition, however, it is a mathematical rather than a physical concept. Closure temperature will vary not only among different thermochronological systems, but also within any single system as a function of grain size, chemical composition and cooling rate. Moreover, it is strictly applicable only in the simplifying case of monotonic cooling. For a more general solution, or to deal with more complicated histories, the equations describing the accumulation and loss of daughter product need to be solved numerically. In Chapter 2, we review the basics of solid-state diffusion, derive an analytical mathematical description of the closure temperature that illustrates the dependence on cooling rate, and present a numerical method to resolve the isotopic age equation for systems that suffer diffusive loss of the daughter product.

The effective closure temperature of any isotopic system will determine what type of process can be addressed by its study: the time and length scales of constraint will scale directly with the closure temperature so that, in any region, higher-temperature systems will constrain deeper, larger-scale processes than will lower-temperature systems. Within the context of this book, we focus on low-to-intermediate-temperature systems, with closure temperatures ≤ 350–$400\,°C$, that record the exhumation history of rocks from upper to mid-crustal levels. Three systems will be considered in detail:

- Ar-based dating methods, utilizing the decay of ^{40}K to ^{40}Ar;
- the (U–Th)/He system that exploits the α-decay of $^{(235,238)}U$ and ^{232}Th; and
- fission-track thermochronology, which is based on the analysis of damage zones in the mineral lattice that result from fission of ^{238}U.

Table 1.1. *Estimated closure temperatures for some commonly used thermochronometers*

Method	Mineral	Closure Temperature (°C)	Reference
K–Ar	Hornblende	500 ± 50	Harrison (1981)
K–Ar	Muscovite	350 ± 50	Hames and Bowring (1994)
K–Ar	Biotite	300 ± 50	Harrison et al. (1985)
K–Ar	K-feldspar	150 – 350	Lovera et al. (1989)
(U–Th)/He	Zircon	200 – 230	Reiners et al. (2002)
(U–Th)/He	Titanite	150 – 200	Reiners and Farley (1999)
(U–Th)/He	Apatite	75 ± 5	Wolf et al. (1998)
Fission track	Zircon	240 ± 20	Brandon et al. (1998)
Fission track	Titanite	265 – 310	Coyle and Wagner (1998)
Fission track	Apatite	110 ± 10	Gleadow and Duddy (1981)

The theoretical and practical bases of these methods are treated in Chapter 3, together with the currently available constraints on their thermal sensitivity. First-order estimates of the closure temperatures for these systems in various minerals are presented in Table 1.1.

There are three possible ways to retrieve information on cooling paths from thermochronological data. The kinetics of ^{40}Ar–^{39}Ar diffusion in feldspar and fission-track annealing in apatite are sufficiently well understood that inverse numerical modelling techniques can give direct constraints on the thermal history from a single sample for these systems. These numerical models will be examined in Chapter 3. Other isotopic dating systems lack sufficiently detailed characterisation to allow such modelling, although progress is being made for the zircon fission-track system as well as for the (U–Th)/He system in apatite.

Where the simplifying assumptions of monotonic and relatively rapid cooling are justified, thermochronological data can be interpreted through the elementary closure-temperature model, as discussed in Section 2.4. In these cases, a thermal history can be retrieved by combining the ages obtained by applying a range of thermochronological systems (with correspondingly varied closure temperatures) to a single sample. Further tectonic and structural insight can also be derived by comparing the ages of multiple samples in close, well-constrained geographic proximity to each other. This approach is commonly applied along local sample transects spanning a large range in altitude (often referred to as 'vertical transects'), and serves to assess at what times samples that are now at different elevations cooled through the closure temperature of the system being studied. As

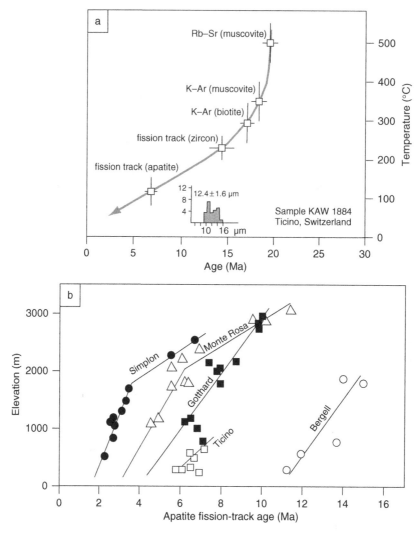

Fig. 1.4. An example of the use of different strategies to obtain time–temperature history data. (a) Multiple-method thermochronology performed on a single sample from the Ticino region, European Alps. Ages (with errors) are plotted against the best estimates of closure temperature for various thermochronometers. Also shown is the apatite fission-track length distribution of the sample. (b) Apatite fission-track age plotted versus elevation for a series of profiles from different locations in the European Alps. Modified from Hurford (1991). Reproduced with permission from Springer-Verlag.

an example, Figure 1.4 (after Hurford (1991)) shows a time–temperature (cooling) path obtained using multiple thermochronometers on a single sample from the European Alps, as well as results from various apatite fission-track age–elevation profiles in the European Alps. Note that these approaches are strictly valid only when thermochronological ages can be interpreted as reflecting the time at which the sample passed through its closure temperature (i.e. monotonic cooling). Where samples have experienced more complex thermal histories, numerical diffusion or annealing modelling approaches are required when one wants to assess the measured ages quantitatively.

1.2 Cooling, denudation and uplift paths

Thermochronology provides information only on the thermal history of rocks, which, among other processes, depends on the rate at which rocks are brought towards the surface by exhumation. Fully extracting the tectono-geomorphic history, i.e., the rate at which rocks are exhumed to the surface by the combined processes of tectonic rock uplift and surface erosion, from the time–temperature results obtained using multiple methods requires an understanding of the interplay among the various physical processes that determine the thermal structure of the lithosphere. These processes are, in turn, related to the various ways by which heat is transported and produced in the crust, namely by conduction, tectonic advection and the decay of radioactive elements. This aspect of interpreting thermochronological data forms the core of this book and is treated in Chapters 4 and 5.

The altitudinal profiling technique has the advantage that no a-priori knowledge of the thermal structure of the crust is required, because the ages of different samples relate to the time of their passage through the same isotherm or thermal interval. However, because such 'vertical' profiles in nature are never truly vertical (except in the case of samples retrieved from drillholes, e.g., Brown *et al.* (2002)), the steepness of the observed age–elevation gradient will be influenced by topographic controls on near-surface thermal structure. Geothermal gradients are also influenced by the rate at which rocks are advected towards the surface, which again may vary regionally. Together with introducing further complexity into the basic interpretation of thermochronological ages, this phenomenon also provides an opportunity to constrain the relief development of a study area, as discussed in Chapters 6 and 7.

As illustrated by Figure 1.4, either the observed thermochronological age can increase more or less linearly with elevation or there can be a distinct break in slope in the age–elevation relationship. Such breaks in slope result from the non-linear increase of annealing or diffusion rates with temperature and reflect the exhumation of partially reset samples. In the limiting case of a stable crustal block (no denudation), a characteristic age profile will develop, whereby rocks situated above the partial-retention zone (PRZ) retain information about an earlier tectonic episode during which they traversed their closure temperature. Their age provides a minimum estimate of the length of the stable period. At depths greater than that of the closure-temperature isotherm, thermochronological ages are effectively zero and there is a zone of transition (the PRZ) wherein ages decrease rapidly with depth. Examples of such profiles of thermochronological age with depth are given in Figures 3.8 and 3.10. When such a crustal block is exhumed rapidly, the PRZ may be preserved as an exhumed or 'fossil' PRZ; its upper and lower boundaries will be expressed as convex-up and concave-up breaks in slope in the age–elevation relationship, respectively.

The key temporal information that is contained in datasets comprising a fossil PRZ is the timing of initiation of the most recent tectonic phase and/or the timing of the end of the previous one. Since the breaks in slope indicate the top and/or base of the fossil PRZ in an age–elevation profile, they can serve as valuable paleo-depth markers (if the geothermal gradient can be quantified) and therefore as markers from which to reconstruct relative vertical offsets (Brown, 1991; Brown *et al.*, 1994; Fitzgerald *et al.*, 1995). The concept of a fossil PRZ has been applied most successfully to the interpretation of apatite fission-track age–elevation profiles (e.g., Fitzgerald *et al.*, 1995, 1999), since the amount of partial annealing of the samples can be monitored through the confined track-length distributions (cf. Section 3.3). The same direct inference cannot be made for isotopic systems where we have only the bulk age at our disposal, but similar patterns have also been observed in (U–Th)/He age–elevation profiles and the conceptual model derived from the fission-track system can be applied to these (House *et al.*, 1997; Stockli *et al.*, 2000; Pik *et al.*, 2003).

Relative depths of exhumation may also be inferred from the spatial variation of thermochronological ages by employing the PRZ concept in the case of geographically distributed samples, in systems for which the depth of exhumation varies spatially (e.g., Tippett and Kamp, 1993; van der Beek *et al.*, 1994; Brandon *et al.*, 1998; Batt *et al.*, 2000). Such systems are, however, also commonly affected by important lateral advection of material (e.g., Batt *et al.*, 2001; Batt and Brandon, 2002), the effects of which on the spatial patterns of thermochronological age are discussed in Chapter 10.

At the other end of the conceptual spectrum, for prolonged constant rates of exhumation under thermal steady-state conditions, thermochronological ages should increase linearly with elevation. If the spatial relationships among the samples permit the problem to be reduced to one dimension (i.e. a truly vertical profile or horizontal isotherms) the slope of the age–elevation relationship will be equal to the denudation rate. As noted above, however, in most cases the sampling profile is not vertical and the isotherms will reflect the surface topography to some degree, in which case the age–elevation gradient will overestimate the true denudation rate. No statement can be made from linear age–elevation trends

1.2 Cooling, denudation and uplift paths

about the total amount of exhumation and the timing of its onset, except that they must be sufficiently large to expose samples from below the PRZ at even the highest elevations. Because of the continuous exhumation in such settings, the system continuously loses its 'memory' as rocks at the surface are eroded and transported away. In order to investigate the evolution of such systems at earlier times than the currently active stage, one may choose to study the thermochronological record of their erosional products rather than investigating the *in situ* thermochronological data. Such *detrital* thermochronological studies, which provide additonal insight but also come with their own set of limitations, are examined in Chapter 9.

In practice (e.g., Figure 1.4), the denudation history will in most cases lie somewhere between these two end-member cases and the observed age–elevation relationship will be a corresponding blend of the two archetypes discussed above. To illustrate what types of relationships may be expected, Figure 1.5 shows predicted apatite fission-track age–elevation trends for samples collected along a 3-km elevation profile, for different denudation histories, ranging from rapid exhumation of a previously stable block to prolonged exhumation at a constant rate.

There has been much confusion in the literature, especially during the 1980s and early 1990s, as to how the thermochronological record can be related to relative vertical motions and offsets of samples. Several review papers have appeared since, with the aim of providing a suitable definition scheme (England and Molnar, 1990; Brown, 1991; Summerfield and Brown, 1998; Ring *et al.*, 1999), which is represented in Figure 1.6. For a set of spatially connected samples in which a fossil PRZ can be identified, the amount of *exhumation*, that is, the upward movement of a rock particle with respect to the surface, is given by the difference in elevation between the bases (or tops) of the fossil and the present-day PRZ, assuming that the geothermal gradient is known and has not changed over time. Note that, although the term *exhumation* has been reserved by the geomorphological community for the strictly limited case of the uncovering and exposure of previously buried elements, such as sub-aerial erosion surfaces (Summerfield and Brown, 1998), we adhere to the broader convention suggested by Ring *et al.* (1999), in which exhumation refers to the progressive exposure of any material particle, irrespective of its prior history. Adopting Ring *et al.* (1999)'s suggested usage, *exhumation* relates to the unroofing history of an actual rock, defined as the vertical distance traversed relative to the Earth's surface, whereas *denudation* relates to the removal of material at a particular point at or under the Earth's surface, by tectonic processes and/or erosion, and is more correctly viewed as a measure of material flux. The two terms can be regarded as synonymous only if effectively one-dimensional behaviour can be assumed,

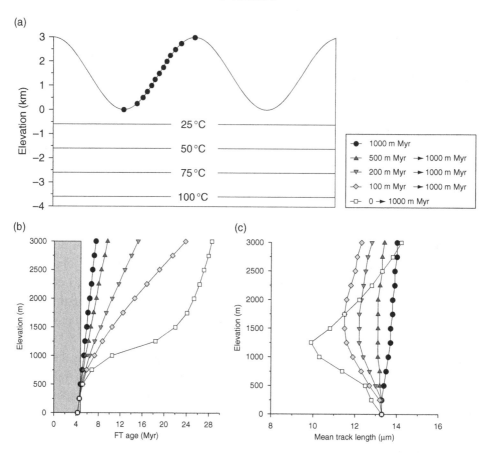

Fig. 1.5. Predicted apatite fission-track age–elevation profiles for different denudation histories. (a) The simplified thermal structure used in the calculations: a geothermal gradient of 25 °C km^{-1} and a surface temperature at sea level of 10 °C. Note that the constant-geotherm approximation is a severe simplification; more realistic models will be shown in Chapters 6 and 7. Black dots on the surface indicate samples for which ages and lengths are predicted. (b) and (c) Predicted apatite fission-track ages and mean track lengths as functions of elevation for five different denudation histories. All histories contain a rapid late denudation phase since 4.8 Myr ago (indicated by the shaded box in (b)) during which material is exhumed at a rate of 1 km Myr^{-1}. Rates previous to 4.8 Myr ago are indicated in the key in the top right-hand corner. Ages and track lengths were predicted using MadTrax code (cf. Appendix 1) using annealing parameters from Laslett *et al.* (1987).

either because regional denudation rates are uniform, or because we are looking at the history of a single sample that has not experienced lateral motion during its exhumation (see Chapter 10).

The amount of *rock uplift*, that is, the upward movement of a rock particle with respect to an external reference frame such as sea level, is equal to the amount of

1.2 Cooling, denudation and uplift paths

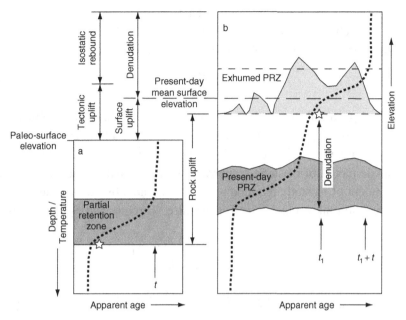

Fig. 1.6. Relationships among different types of uplift, denudation and thermochronological age–elevation pattern. (a) The initial thermochronological age–depth profile, established under conditions of relative tectonic stability during a time t. Thermochronological ages decrease rapidly in the partial-retention zone (PRZ) (or partial annealing zone for fission tracks) as in Figure 3.8. The depth to the base (or top) of the PRZ depends on the geothermal gradient. (b) At time t_1, the crustal section starts to be uplifted and partially eroded. If the pre-existing PRZ is preserved, a sample from its base (marked by a star) will record the age of onset of denudation t_1 and its elevation with respect to its original depth will equal the amount of rock uplift. Note that, in order to reconstruct the amount of tectonic, surface and rock uplift, the initial surface elevation at t_1 needs to be reconstructed. Without this constraint, only the amount of denudation can be quantified, from the elevation of the base of the fossil PRZ. Modified from Fitzgerald *et al.* (1995). Reproduced with permission from the American Geophysical Union.

denudation plus the amount of *surface uplift*, the uplift of the surface with respect to that reference frame. Quantifying the latter requires an independent estimate of the paleo-surface elevation to be made, which is possible only in rare cases (e.g., Abbott *et al.*, 1997). In the context of this textbook, we consider a range of tectonic settings, ranging from situations where rocks are actively being uplifted by ongoing tectonic processes and erosion counteracts the resulting uplift of the surface (Chapter 13) to situations where, in the absence of any tectonic forcing, rock uplift is driven solely by the isostatic rebound caused by erosional unloading (Chapters 11 and 12).

The lithosphere will respond to denudational unloading by isostatic rebound (Turcotte and Schubert, 1982); in the simplest case of local isostasy the amount of isostatic rebound I will be related to the amount of denudation E by

$$I = \frac{\rho_c}{\rho_m} E \qquad (1.1)$$

where ρ_c and ρ_m are the crustal and mantle densities, respectively. Since typical values for ρ_c and ρ_m are of the order of 2750 and 3300 kg m^{-3}, respectively, isostatic rebound may equal up to five-sixths of the amount of denudation. A more realistic model of isostatic response is flexural isostasy, however, in which case the amount of isostatic rebound will be modulated by the wavelength λ over which denudation takes place (see, for instance, Turcotte and Schubert, 1982):

$$I = \frac{E}{\frac{\rho_m}{\rho_c} - 1 + \frac{D}{\rho_c g}\left(\frac{2\pi}{\lambda}\right)^4} \qquad (1.2)$$

where g is the acceleration due to gravity and D is the flexural rigidity of the lithosphere, given by

$$D = \frac{Y_m T_e^3}{12(1-\nu^2)} \qquad (1.3)$$

where Y_m is Young's modulus, T_e is the effective elastic thickness of the lithosphere and ν is Pascal's ratio (cf. Section 11.2). The quantity $\lambda_c = [D/\rho_c g]^{1/4}$ has the dimensions of a length and is commonly termed the 'natural flexural wavelength' of the lithosphere (Turcotte and Schubert, 1982). Erosional unloading that takes place on a wavelength much shorter than λ_c will not be accompanied by a substantial amount of isostatic rebound; erosional unloading on a wavelength greater than λ_c will be fully isostatically compensated. The interplay between denudation and the isostatic response of the lithosphere, as well as its consequences for thermochronology, are more fully discussed in Chapter 11.

Finally, a measure of particular interest is the amount of *tectonic uplift* (U_T), that is, the amount of rock uplift (U_r) that is not accounted for by isostatic rebound. From Figure 1.6, it follows that

$$U_T = U_r - I = h_0 - h_i + E\left(1 - \frac{\rho_c}{\rho_m}\right) \qquad (1.4)$$

under the assumption of local isostasy. h_0 and h_i are the present-day and initial surface elevations, respectively. From the above, it is clear that any absolute measure of uplift (whether it be rock, surface, or tectonic uplift) requires knowledge of the initial surface elevation, a measure that cannot be obtained with thermochronological methods.

1.3 Thermochronology in practice

The optimal sampling strategy to apply to a given thermochronological problem – which thermochronological system to use, where and how to sample – will always be a compromise between the research ideals and the practicalities of local geology.

Choice of thermochronometer

The precise sensitivity of any chronometer in a particular situation is a complex variable, integrating the temperatures of open- and closed-system behaviours, the rates of exhumation experienced, local thermal structure and the length scale over which deformation occurs: there is no simple hierarchy that governs which chronometers will be best suited to resolve particular styles of problem. The relative behaviour of a range of thermochronometers can, however, be used to set out a range of general principles guiding the planning and execution of an optimal thermo-tectonic study, particularly where reconnaissance data have already established the basic response of one or more thermochronological systems in an area.

The higher the closure temperature concerned, the deeper a chronometer undergoes closure for a given exhumation path and the longer a sample takes to be exhumed following closure (i.e. the greater its age upon exposure at the surface). It follows that, in general terms, no chronometer is uniquely suited for a particular task, but rather closure of different thermochronometers at differing temperatures results in sensitivity to a corresponding range of time and length scales of behaviour. The higher the closure temperature of a chronometer, the more sensitive its ages will be to the variations in thermal structure and denudation rate across a region and to the specific exhumation path experienced. Conversely, the lower the closure temperature, the more sensitive a chronometer will be to short-term behaviour and local conditions.

Sampling scale

Two principal assumptions underlie the elementary approach of interpreting cooling ages in terms of *rates* of tectonic processes. (1) Samples must be exhumed from below the depth at which the ambient crustal temperature is equal to the closure temperature of the chronometer in question. Although isotopic diffusion is not in itself dependent on pressure, the importance of subsequent exhumation to the final disposition of ages at the surface makes the depth at which closure occurs (the closure depth) a fundamentally important quantity in the tectonic interpretation of thermochronology. (2) The isotopic ages observed should be at steady-state values, such that they reflect the time taken for a sample to be exhumed from its closure depth. If this requirement is not fulfilled, ages reflect an average of the exhumation rates experienced between closure and exhumation, and so provide no direct information about local dynamics at any particular point in time. The more comprehensive interpretation of these complex isotopic age records is discussed in greater depth in Chapter 5 and in Section 13.3. The second assumption also requires that the region concerned be in a topographic and thermal steady state on the timescales over which chronometers are exhumed following closure, since a change in either of these characteristics can alter the relevant closure depth, bringing about an accompanying transience in cooling ages at the surface (see Chapters 5 and 6).

To investigate the thermo-tectonic evolution of a given orogenic system fully, the spatial scale of constraint should optimally be comparable to the length scale of the thermal structure. Thermal perturbations to the geothermal structure, for example, vary in proportion with relief, since the thermal disturbance produced by topography will decay exponentially over a depth range equal to the wavelength of the topography (Turcotte and Schubert, 1982; Stüwe *et al.*, 1994; Mancktelow and Grasemann, 1997; Braun, 2002a). While the low-temperature constraint of (U–Th)/He in apatite may thus potentially reveal insight into surface processes, such constraint is generally beyond the potential of higher-temperature systems, for which the thermal consequences of relative relief development and modification usually do not penetrate to the relevant isotherm depth in the crust, except for exceedingly high exhumation rates.

Structural complexity

Because geographically separated sample transects and spatial trends in data invariably play a role in interpreting the physical implications of thermochronological age, mapping and consideration of the effects of structural features should be considered. If structural offset pre-dates or occurs at a deeper level in the crust than the closure of the relevant chronometer, it should have no influence on the ages observed. In contrast, if structures experience significant offset during the post-closure interval, this can result in tilting or offset of the age distribution at the surface, requiring correction before the data can be meaningfully interpreted.

Mineralogy

The potential for structured investigation of a specific problem is fundamentally limited by the distribution of lithologies in the study area. Local geological character controls the availability of various chronometers in a given setting,

1.3 Thermochronology in practice

and exposure is not universal or guaranteed in otherwise optimal sampling locations. A good example of this is provided by the distribution of datable phases in the Southern Alps of New Zealand. The geology of this orogen is relatively uniform, with the region dominated almost to exclusion by a sequence of predominantly metasedimentary schists and gneisses. The metamorphic grade of this suite increases essentially continuously from lower greenschist facies in the Main Divide of the Southern Alps to a maximum of lower amphibolite facies adjacent to the Alpine Fault (Figure 1.7).

Apatite, zircon, sphene and other accessory U- and Th-bearing phases are present throughout this region. Although these allow widespread fission-track dating, the abraded and variably corroded character of the recycled sedimentary grains leaves them rather poorly suited for the detailed optical characterisation required for (U–Th)/He dating – particularly for apatite. Higher-temperature thermochronometers are far more restricted in their distribution. The first appearance of prograde biotite occurs between 15 and 25 km from the Alpine Fault (Mason, 1962). This biotite may initially be present as microscopic flakes, but is more commonly observed as porphyroblasts 1 mm or more in diameter. Although relict

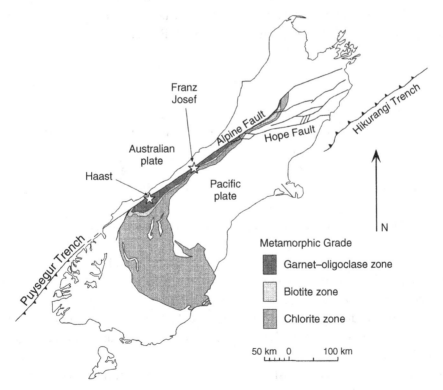

Fig. 1.7. The distribution of metamorphic rocks in the Southern Alps, South Island, New Zealand.

sedimentary muscovite and fine-grained serricite occur throughout much of the metamorphic sequence, this again is routinely separable only once it develops a porphyroblastic character within the biotite zone. This restricted distribution proved a limitation in the K–Ar and ^{40}Ar–^{39}Ar study of Batt *et al.* (2000), where the transition between major behavioural domains in the Ar dataset was inferred to occur at or beyond the limits of separable mica, preventing the desired level of behavioural constraint of the orogen.

Prograde metamorphic hornblende comes into the sequence even further west in the garnet zone, and even then is largely limited to the metabasic schist members (Mason, 1962). This monotonic metamorphic sequence is broken only by a volumetrically minor suite of granitic pegmatite bodies that crop out within 2–3 km of the Alpine Fault in the Mataketake Range, Mount Kinnaird, and the Paringa River valley in the south of the orogen (Figure 1.7) (Wallace, 1974; Chamberlain *et al.*, 1995; Batt and Braun, 1999), and a series of lamprophyric dykes that occur in this area and further to the southeast (Adams and Cooper, 1996). Despite the limited volume and extent of these lithologies, their amenability to varied isotopic dating approaches (with members rich in biotite, muscovite, hornblende and even K-feldspar) has seen them assume disproportionate significance in studies of the evolution of the Southern Alps on longer timescales (Chamberlain *et al.*, 1995; Adams and Cooper, 1996; Batt and Braun, 1999; Batt *et al.*, 2000).

Sample preparation

In preparing a purified mineral separate, complete disaggregation of the rock is a required first step. For bulk mineral phases, this inevitably means comminution of grains and reduction of grain size.

The principal concern in this process is over-crushing. While it is less significant for fission-track analysis, in which the grain interior is exposed as a matter of standard protocol, reduction of grain size can be a problem for isotopic methods relying on diffusive behaviour and retention of radiogenic daughter products.

In many cases it has been demonstrated that sub-grain-scale features (cleavage, linear dislocation features, cracks etc.) control the diffusion dimension. If a sample is crushed below this size, the fundamental controls on radiogenic-gas (i.e. Ar or He) retention during the sample's geological history will not be reflected in the gas release observed in the laboratory, and constraining the relevant diffusion properties will not be possible.

It has been found that diffusive release of argon from hornblende, biotite and feldspar is not affected by crushing to 100 μm, demonstrating that the natural dimension controlling gas retention and loss is below that size (Harrison, 1981;

1.3 Thermochronology in practice

Harrison et al., 1985; Lovera et al., 1989; Foster et al., 1990). Using this limit as the lower cut-off in size of the crushed fraction allows relatively simple and efficient separation of material by density and magnetic separation, and grain-crushing effects are not commonly considered as a factor in the interpretation of such ages.

Crushing is of immediate concern in (U–Th)/He dating, however, due to the significance of numerical correction for the effects of long α-particle stopping distances and consequent ejection from grain margins (see Section 3.2). If the grain exterior over the dimensions to which this correction is applied ($\sim 20\,\mu m$) is partially removed, the correction will over-compensate, resulting in a spurious reduction in age.

Sample quality

Basic sample character is not often stressed in discussions of thermochronological data, but may be very important in controlling data quality and reproducibility.

The U- and Th-content of the rocks may pose a problem for fission-track analyses, since the abundance of parent isotopes determines the quantity of fission tracks produced. The U-content in 'young' (less than a few Myr) apatites or titanites may be too low for one to obtain precise ages because grains with no fission tracks are common. In contrast, in zircons tens to hundreds of Myr old, the U-content, and its associated α-damage, may be so large that the grains are metamictised – they turn black upon chemical etching and cannot be dated using the fission-track technique.

In (U–Th)/He dating of apatite (Section 3.2), the focus on sample character is particularly acute. Successful implementation of the standard experimental protocols for this method depends on the ability to pick good-quality, inclusion-free grains, and many workers have experienced difficulties involving data reproducibility because of this issue.

The numerical correction for α-particle ejection relies on grain morphology approximating an idealised crystal model, such that uniform, euhedral grains are preferred. In multi-grain aliquots, all grains are weighted equally in the calculation of mean dimensions. As long as the grain population has uniform U and Th content, this assumption has no manifest negative consequences, but if this content varies between grains, then those with higher U and Th contents will contribute proportionately more to the overall helium produced, and hence dominate the relative ejection-loss statisitics for the sample. In the absence of a-priori knowledge of the relative U and Th contents of the individual grains in a sample aliquot, it follows that it is best practice to minimize the size variation between grains, so that all should have a consistent α-ejection profile in any case.

While fluid inclusions are generally avoided for all samples, because of their potential to contain parent or daughter isotopes from an external reservoir as a solute phase, the presence of crystalline inclusions also requires extra care for apatite. Apatite bears relatively little U and Th in comparison with other accessory phases such as zircon and sphene. If inclusions of these other, more refractory minerals are present within an apatite, it would affect the dating process and calculations in two ways.

(i) Because the α-recoil distance is approximately 20 μm, inclusions smaller than this dimension will not retain significant quantities of the He they produce, and their diffusional properties should thus not influence the experimental release of He during heating. However, refractory inclusions such as zircon are not dissolved by the standard techniques used with apatite. The helium they contribute to the system will thus be unsupported by U and Th in the analysis, leading to an older apparent age.
(ii) The standard means of thermochronological interpretation of ages use modelling of diffusion and α-recoil within the crystal. If the distribution of helium is inhomogeneous, this modelling approach will be invalid.

Owing to these issues of sample character and the demanding standards they impose on sample quality, it is standard practice with most thermochronometric methods to run replicate samples in order to assess the reproducibility of the results obtained. The significance of this procedure is confirmed by the observation that age variations between replicates are commonly larger than the theoretical error of the analytical techniques applied.

2
Basics of thermochronology: from t–T paths to ages

The processes underlying thermochronology are the accumulation of the daughter products of radioactive decay and their subsequent diffusion through and out of the host mineral. The accumulation of daughter products (which determines the thermochronological age of a sample) results from the interplay between these two processes. Since diffusion rates are exponentially dependent on temperature, there exists a highly non-linear relationship between the thermal history a sample has experienced and its thermochronological age. In this chapter, we describe how one can predict accumulation of daughter products from a given thermal history. This is a very important step that links the predictions of a given tectono-morphic scenario to observables (ages). Various methods are proposed, depending on the level of accuracy with which one aims to describe the physical process. The discussion is cast in terms of isotopic systems; the same general principles hold, however, for thermochronological dating techniques (such as fission-track dating) based on the accumulation of damage zones in the crystal lattice.

2.1 The isotopic age equation

The basic principle behind most thermochronometric methods is that an isotope (the daughter isotope) is produced by the radioactive decay of a naturally occurring unstable isotope (the parent isotope) present in the rock. The age constraint provided by a thermochronometer is based on this decay at a known rate, that is proportional only to the number of parent atoms (N_p) left at any time:

$$\frac{dN_p}{dt} = -\lambda N_p \tag{2.1}$$

where λ is known as the decay constant and has units of time^{-1}. The change in parent-isotope abundance over an interval of time is found by integrating (2.1) and setting the original abundance $(N_p)_0$:

$$(N_p)_t = (N_p)_0 e^{-\lambda t} \tag{2.2}$$

The rate of decay of a radioactive isotope is also commonly expressed in terms of its half-life $t_{1/2}$, defined as the time it takes for half the original amount of parent atoms to decay; $N_p = \frac{1}{2}(N_p)_0$, so that $t_{1/2} = \ln 2/\lambda$. Given constraints on $(N_p)_0$ and the abundance of daughter atoms N_d, Equation (2.2) can be solved for time in order to obtain a radiometric age. However, the original abundance of the parent isotope is a fundamental unknown. Therefore, the abundances of both the parent and the daughter isotopes N_p and N_d are usually measured in order to calculate an isotopic age for a sample, assuming the operation of a closed system (i.e. no loss of daughter isotope). Since

$$(N_d)_t = (N_p)_0 - (N_p)_t \tag{2.3}$$

we can write, by solving (2.2) for $(N_p)_0$ and combining the result with (2.3),

$$N_d = N_p(e^{\lambda t} - 1) \tag{2.4}$$

which leads to the general form of the isotopic age equation:

$$t = \frac{1}{\lambda} \ln\left(\frac{N_d}{N_p} + 1\right) \tag{2.5}$$

For several isotopes important in isotopic dating, decay is possible by any one of several different pathways, producing a corresponding variety in potential daughter products. For example, ^{40}K can decay to ^{40}Ar either by electron capture or by β^+ decay, or to ^{40}Ca by β^- decay. ^{238}U can decay either by spontaneous fission or in a series of α and β^- emissions. In the general case in which two daughter isotopes N_{d_1} and N_{d_2} are produced, with decay constants λ_1 and λ_2, the total decay constant $\lambda = \lambda_1 + \lambda_2$ is introduced and

$$N_{d_1} = \frac{\lambda_1}{\lambda} N_p(e^{\lambda t} - 1) \tag{2.6}$$

which can be solved for time to give

$$t = \frac{1}{\lambda} \ln\left(\frac{\lambda}{\lambda_1} \frac{N_{d_1}}{N_p} + 1\right) \tag{2.7}$$

2.2 Solid-state diffusion – the basic equation

The daughter isotope is a chemical species that is foreign to the host mineral and might not be suited to retention in the crystal lattice. Non-reactive daughter products will therefore diffuse through the host mineral over time. As they reach the mineral surface, they become subject to rapid fluid transport and are effectively 'lost' from the system, leading to a negligible concentration of daughter isotopes on the mineral surface. The resulting concentration gradient within the host mineral grain drives diffusion of daughter isotopes out of the system. More reactive decay products, such as Sr and Pb, may more easily be retained in the host mineral or the surrounding phases, suppressing diffusion out of the system.

Temperature exerts a first-order control on the rate of diffusive processes, which can be parameterised by a simple diffusion equation:

$$\frac{\partial N_d}{\partial t} = D(T)\nabla^2 N_d + P \qquad (2.8)$$

where N_d is the concentration of the daughter element, the ∇^2 symbol represents the general Laplacian operator (the second-order spatial derivative), P is the rate of production of the daughter element and the diffusivity $D(T)$ is a strong function of temperature (T), usually of the Arrhenius type:

$$\frac{D(T)}{a^2} = \frac{D_0}{a^2} e^{-E_a/(RT)} \qquad (2.9)$$

where D_0, the diffusivity at infinite temperature, is known as the diffusion constant, a is the dimension of the diffusion domain, which in simple systems can be the physical size (radius) of the grain, but can also be some sub-grain structure with a correspondingly smaller size, E_a is the activation energy and R is the gas constant. The assumption that the daughter element is rapidly removed from the system once it has reached the grain boundary is equivalent to assuming that, on the grain boundary, the concentration, N_d, is zero. This assumption is an essential element of any system for quantitatively interpreting isotopic age.

For radioactive systems in which the half-life of the decay reaction is much greater than the time over which we integrate the diffusion equation (i.e. the thermochronological age of the sample), the production rate, P, which is proportional to the concentration of the parent element, can be assumed to remain constant through time. In the absence of conflicting evidence, it is also commonly assumed that production is spatially uniform, i.e. the parent element is uniformly distributed within the grain.

As the diffusion parameters, D_0, E_a and a vary for different isotopic species and mineral structures; each geochronometer has its own specific range of temperatures at which the daughter isotopes reduce and eventually cease their mobility within the crystal lattice. For classical geochronological applications, where the interest is in the crystallisation age of a rock or mineral, isotopic methods for which the daughter product is retained at high temperatures, such as U–Pb, Sm–Nd and Rb–Sr, are most useful, since the apparent age of these chronometers does not depend strongly on their subsequent thermal history.

A thermochronometer, in contrast, provides an apparent thermal age for a rock, which for simple cooling histories can be thought of as the time in the past when the rock passed through a temperature or range of temperatures below which the daughter product is retained. Three different temperatures that are relevant to the process of increasing isotopic retention can be defined (Figure 2.1). At high temperatures, the diffusivity is so large that the system is completely open: any daughter isotope produced is instantaneously lost. As the rock cools, the diffusivity decreases and some of the daughter product will start to be retained. The temperature at which this happens is the lower limit of fully open-system behaviour, which we will call the *'open-system' temperature* for lack of a more rigorously defined term. Upon further cooling, the diffusivity decreases further until diffusion becomes so sluggish that all newly produced daughter isotopes are effectively retained in the host crystal on geological timescales: the system

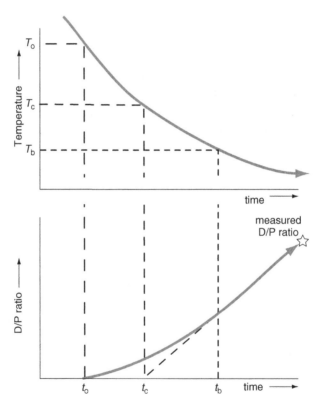

Fig. 2.1. Definitions of the 'open-system', closure and blocking temperatures (T_o, T_c and T_b, respectively) in a cooling thermochronometric system. The upper plot schematically shows the evolution of temperature through time; the lower plot the evolution of the ratio of daughter to parent isotope. t_c is the apparent thermochronological age of the system; t_o and t_b correspond to the times of initial (partial) retention (cessation of fully open-system behaviour) and blocking (initiation of closed-system behaviour), respectively. Modified from Dodson (1973).

has become closed. The upper temperature limit of closed-system behaviour is known as the *blocking temperature*; within the temperature interval between the blocking and the open-system temperatures *partial* retention occurs. In a spatial reference frame, the depth interval between the isotherms corresponding to the blocking and open-system temperatures is known as the *partial-retention zone*. Although both the open-system and blocking temperatures have a precise definition, they are practically impossible to constrain directly. In practice, we will measure a daughter-to-parent ratio in the host mineral (indicated by the star in Figure 2.1) that has built up during the phase of closed-system behaviour as well as during the preceding phase of partial retention. On applying Equation (2.5), we find an apparent thermochronological age t_c. Dodson (1973) introduced the term *closure temperature*, which he defined as the temperature of the thermochronological system at the time t_c corresponding to its apparent age (cf. Figure 2.1). The closure temperature, which by definition lies between the open-system and blocking temperatures, provides the simplest conceptual entity to which a thermochronological age can be related. The significance of any isotopic age with respect to the geological history of a sample, however, depends explicitly on the thermal history experienced relative to the key temperature interval that is the partial-retention zone.

2.3 Absolute closure-temperature approximation

Because the temperature dependence of diffusivity is exponential (see Equation (2.9)), one can make the simple, first-order approximation that, for each thermochronological system, the daughter product is completely lost above the closure temperature and perfectly stored in the host mineral below it. This simplifying assumption boils down to collapsing the partial-retention zone, and its upper and lower boundaries defined by the open-system and blocking temperatures, respectively, onto the closure temperature. Approximate values for the closure temperature, based either on extrapolation of laboratory measurements of diffusivity or on field-based studies in which the temperature histories of the samples are well constrained, are given for a range of thermochronometers in Table 1.1. It should be noted that, in a strict sense, this concept of the closure temperature can be introduced only for systems that undergo simple linear cooling.

The simplest approach to predicting thermochronological ages from a given temperature history is thus to state that the age is the effective time at which the rock cooled below the closure temperature. However, the closure temperature depends not only on the diffusion properties of the daughter product in a given mineral but also on the geometry and size of the grain, as well as on the cooling rate. A more rigorous, even though empirical, approach was proposed by Dodson (1973) in order to take these variables into account.

2.4 Dodson's method

Dodson (1973) introduced the notion of the *closure temperature*, T_c, as the temperature of a thermochronological system at the time corresponding to its apparent age and demonstrated that, for a given set of diffusion parameters, D_0/a^2 and E_a, the closure temperature depends on the cooling rate. Conceptually, we can understand this because the cooling rate determines the time the sample will spend in the partial-retention zone and therefore the amount of daughter product that will build up during this transitional interval (Figure 2.2). Since the closure temperature is defined with respect to the apparent age (and thus the measured concentration of the daughter isotope) it should also vary with cooling rate. Dodson (1973) provided an elegant mathematical model to demonstrate this behaviour.

Under the assumption that the temperature of a rock varies as the inverse of time as it passes the partial-retention zone, $T(t) \propto 1/t$, one can find an approximate analytical solution to Equation (2.8). In this situation, the diffusivity, D, varies exponentially with time:

$$D(t) = D(0)e^{-t/\tau} \tag{2.10}$$

where τ is the time taken for the diffusivity to decrease by a factor e. τ can be related to the cooling rate, $\dot{T}$, by noting the relationship

$$\frac{\partial E_a/(RT)}{\partial t} = \frac{1}{\tau}$$

$$\tau = \frac{R}{E_a \, \partial T^{-1}/\partial t}$$

$$= -\frac{RT^2}{E_a \dot{T}} \tag{2.11}$$

By convention, because the rock is cooling, $\dot{T}$ is negative.

The evolution equation (2.8) for the concentration of the daughter product, N_d, becomes

$$\frac{\partial N_d/N_p}{\partial \theta} = \frac{\tau D(0)}{a^2} e^{-\theta} \nabla^2 (N_d/N_p) + P/N_p \tag{2.12}$$

where N_p is the parent concentration and θ is dimensionless time, $\theta = t/\tau$. N_d/N_p is thus a function of $\tau D(0)/a^2$, θ and P. At very large times, the system is cold enough that all daughter isotopes are stored in the grain and N_d/N_p increases

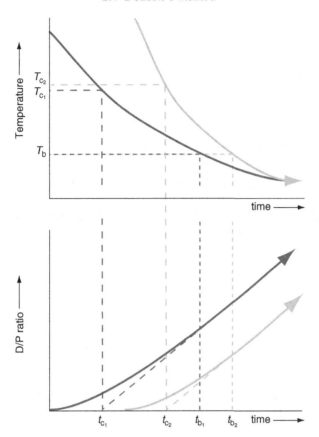

Fig. 2.2. A comparison of the buildup of thermochronological 'age' for slowly cooling (dark shaded) and rapidly cooling (light shaded) samples. T_{c_1} and T_{c_2} are the closure temperatures for the slowly and rapidly cooling samples, respectively, and t_{c_1} and t_{c_2} are their respective apparent ages. T_b is the blocking temperature that is independent of cooling rate; t_{b_1} and t_{b_2} are the blocking times for the slowly and rapidly cooling samples, respectively. The figure shows that the longer the interval of partial retention (i.e. the slower the cooling), the further apparent closure occurs from blocking, and the lower the closure temperature. Note that the difference in cooling temperatures is exaggerated in this figure; because of the logarithmic dependence of cooling temperature on cooling rates (see Equation (2.21)), the sensitivity of the former to the latter is in fact limited.

linearly with time, $N_d/N_p = f(\theta, P)$. For Dodson, the closure time, θ_c, corresponds to the extrapolation of this relationship to zero (see Figure 2.1). The *closure temperature*, T_c, is the temperature corresponding to the closure time in the rock's thermal history. We can thus write

$$0 = f(\tau D(0)/a^2, \theta_c, P) \qquad (2.13)$$

or

$$\theta_c = g(\tau D(0)/a^2, P) \tag{2.14}$$

From the definition of τ, we can write

$$\theta_c = \frac{E_a}{RT_c} - \frac{E_a}{RT_0} \tag{2.15}$$

where T_0 is the temperature at time $t = 0$, and

$$\frac{E_a}{RT_c} = \frac{E_a}{RT_0} + g(\tau D_0 e^{-E_a/(RT_0)}/a^2, P) \tag{2.16}$$

On introducing the substitute parameters $X = E_a/(RT_0)$ and $u = \tau D_0 e^{-X}/a^2$, we can write the condition that T_c must be independent of T_0 as

$$\frac{\partial E_a/(RT_c)}{\partial X} = 0 \tag{2.17}$$

This leads to

$$0 = 1 + \left.\frac{\partial g}{\partial u}\right|_P \frac{\partial u}{\partial X} = 1 - u \left.\frac{\partial g}{\partial u}\right|_P \tag{2.18}$$

or

$$g = \ln u + b(P) = \ln(\tau D_0/a^2) - E_a/(RT_0) + b(P) \tag{2.19}$$

Finally, one can derive the following expression for the closure temperature T_c:

$$E_a/(RT_c) = \ln(\tau D_0/a^2) + b(P) \tag{2.20}$$

The term b will be different for various assumed geometries and production rates; it is usually incorporated into this expression as a factor A ($A = 25$ for a sphere, 27 for a cylinder and 8.7 for a plane sheet). These geometrical constants arise as a consequence of the comparative surface-area/volume ratios of these different forms. We can solve (2.20) for the closure temperature, T_c:

$$T_c = \frac{E_a}{R \ln(A \tau D_0/a^2)} \tag{2.21}$$

Thus, by writing $T = T_c$ in Equation (2.11) and combining it with (2.21), one can derive the value of the closure temperature from an assumed cooling rate and the values of experimentally determined diffusion coefficients. Because both relationships depend on T_c, an iterative solution must be employed. However, the iterative scheme converges very rapidly on a unique value of the closure temperature.

Equation (2.21) demonstrates the positive dependence of the closure temperature of any given thermochronological system on the rate at which the rock is

2.5 Numerical solution

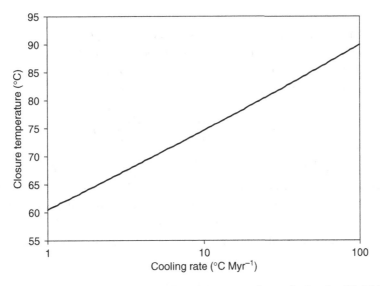

Fig. 2.3. The closure temperature of a 100-μm apatite grain for the (U–Th)/He system. Values of $D_0/a^2 = 50.1 \times 10^6$ s^{-1} and $E_a = 151.46$ kJ mol^{-1} were assumed.

cooled. Figure 2.3 shows this dependence for a particular isotopic system, (U–Th)/He in apatite. It also demonstrates that this dependence, although important for our purposes, is relatively weak: an increase in cooling rate by two orders of magnitude results in a ~10% change in T_c (when expressed as absolute temperatures) in this example.

2.5 Numerical solution

The above analysis of thermochronological ages in terms of cooling through a particular temperature applies only to the limited case of monotonic cooling (and strictly only to the further limited case in which temperature varies as the inverse of time: $T(t) \propto 1/t$). For a more general solution, or to deal with more complicated histories, the solid-state diffusion equation needs to be solved numerically. Numerical solutions of Equation (2.8) imply assumptions on the geometry of the grain, the parent distribution and the diffusion properties of the mineral. A relatively efficient and flexible solution is provided by the finite-difference scheme presented below.

The spherical approximation

Assuming that the diffusion domains are small spherical objects of radius a, Equation (2.8) can be written as a function of the dimensionless radial distance, r, as

$$\frac{\partial N_d}{\partial t} = \frac{D}{a^2}\left(\frac{\partial^2 N_d}{\partial r^2} + \frac{2}{r}\frac{\partial N_d}{\partial r}\right) + P \qquad (2.22)$$

The thermochronological 'age' of a sample is obtained from the ratio of the concentration of the daughter element, N_d, and the production rate P (which depends on the concentration of the parent element, as described in Section 2.1) integrated over the sample volume. One can therefore write the following evolution equation for the apparent age $A(r, t) = N_d(r, t)/P$:

$$\frac{\partial A(r,t)}{\partial t} = \frac{D}{a^2}\left(\frac{\partial^2 A(r,t)}{\partial r^2} + \frac{2}{r}\frac{\partial A(r,t)}{\partial r}\right) + 1 \qquad (2.23)$$

This equation can be integrated through time for a given thermal history $T(t)$, which determines the value of the diffusivity $D(T)$. The initial value $A(r, 0) = A_0$ corresponds to the initial age of the grain, i.e. before the thermal history determined by $T(t)$ begins. The boundary condition is $A(1, t) = 0$. The age of the grain is obtained by spatially averaging $A(r, t)$ over the volume of the grain (see Appendix 6):

$$\text{age} = 3 \int_0^1 r^2 A(r, t) dr \qquad (2.24)$$

At low temperature, the diffusivity is small ($D \to 0$) and the evolution equation (2.23) becomes

$$\frac{\partial A(r,t)}{\partial t} = 1 \qquad (2.25)$$

which shows that, everywhere in the grain, age increases linearly with time. At high temperature, the diffusivity is large ($D \to \infty$) and no daughter isotopes accumulate within the grain ($\partial A(r,t)/\partial t = 0$). Equation (2.23) becomes

$$\frac{\partial^2 A(r,t)}{\partial r^2} + \frac{2}{r}\frac{\partial A(r,t)}{\partial r} = 0 \qquad (2.26)$$

implying that no gradient in age is permitted within the grain.

The finite-difference formulation

Our purpose is to find the solution to (2.23) at a finite number of points along the radius of the spherical grain, $r_i = (i-1)\Delta r$, for $i = 1, \ldots, n$. The solution at r_i is denoted A_i. Derivatives with respect to r are approximated by a centred difference:

$$\frac{\partial A}{\partial r}(r_i) = \frac{A_{i+1} - A_{i-1}}{2\Delta r}$$
$$\frac{\partial^2 A}{\partial r^2}(r_i) = \frac{A_{i+1} - 2A_i + A_{i-1}}{\Delta r^2} \qquad (2.27)$$

2.5 Numerical solution

The time derivative is approximated by

$$\frac{\partial A_i}{\partial t} = \alpha \frac{\partial A_i^{t+\Delta t}}{\partial t} + (1-\alpha) \frac{\partial A_i^t}{\partial t} \qquad (2.28)$$

The factor α will determine whether we use an explicit ($\alpha = 0$) or implicit ($\alpha = 1$) time-marching method. As we will see below, an explicit method is rather efficient because it does not require the solution of a large system of algebraic equations. The equations are not coupled, i.e., each equation has only one unknown. An implicit method leads to a system of coupled equations that require the solution (or inversion) of a tri-diagonal matrix system. Explicit methods can be unstable (i.e., the solution can diverge if the time step is not made sufficiently small); implicit methods are always stable. Both require time steps shorter than a critical length in order to provide an accurate solution. Intermediate values of α correspond to explicit–implicit schemes that are, in general, both stable and accurate (see Belytschko et al. (1979) for a complete description of mixed time-integration methods).

The solution at $t + \Delta t$ is expressed as a Taylor expansion:

$$A_i^{t+\Delta t} = A_i^t + \Delta t \frac{\partial A_i}{\partial t} + O(\Delta t^2) + \cdots \qquad (2.29)$$

in which we neglect the terms in Δt^2 and higher orders to obtain

$$A_i^{t+\Delta t} = A_i^t + \Delta t \left(\alpha \frac{\partial A_i^{t+\Delta t}}{\partial t} + (1-\alpha) \frac{\partial A_i^t}{\partial t} \right) \qquad (2.30)$$

which, using (2.23), leads to the following finite-difference form of the evolution equation:

$$A_i^{t+\Delta t} = A_i^t + \alpha \Delta t \frac{D^{t+\Delta t}}{a^2} \left(\frac{A_{i+1}^{t+\Delta t} - 2A_i^{t+\Delta t} + A_{i-1}^{t+\Delta t}}{\Delta r^2} + \frac{2}{r_i} \frac{A_{i+1}^{t+\Delta t} - A_{i-1}^{t+\Delta t}}{2\Delta r} \right)$$
$$+ (1-\alpha) \Delta t \frac{D^t}{a^2} \left(\frac{A_{i+1}^t - 2A_i^t + A_{i-1}^t}{\Delta r^2} + \frac{2}{r_i} \frac{A_{i+1}^t - A_{i-1}^t}{2\Delta r} \right) + \Delta t \qquad (2.31)$$

We see now that, in the case for which $\alpha = 0$, i.e. the method is explicit, the value of each $A_i^{t+\Delta t}$ can be directly obtained from the values of A^t at $i-1$, i and $i+1$; there is no need to find the solution of a large system of coupled algebraic equations.

When $\alpha \neq 0$, the solution can be obtained only by finding the solution of a tri-diagonal system of algebraic equations. Each equation can be re-written as

$$L_i A_{i-1}^{t+\Delta t} + D_i A_i^{t+\Delta t} + U_i A_{i+1}^{t+\Delta t} = b_i, \qquad \text{for } i = 2, \ldots, n-1 \qquad (2.32)$$

where, noting that $\Delta r/r_i = 1/(i-1)$,

$$U_i = -\frac{\alpha \Delta t D^{t+\Delta t}}{\Delta r^2 a^2}\left(1 + \frac{1}{i-1}\right)$$

$$D_i = 1 + \frac{2\alpha \Delta t D^{t+\Delta t}}{\Delta r^2 a^2}$$

$$L_i = -\frac{\alpha \Delta t D^{t+\Delta t}}{\Delta r^2 a^2}\left(1 - \frac{1}{i-1}\right)$$

$$b_i = \Delta t + \frac{(1-\alpha)\Delta t D^t}{\Delta r^2 a^2}\left[A^t_{i+1}\left(1 + \frac{1}{i-1}\right) - 2A^t_i + A^t_{i-1}\left(1 - \frac{1}{i-1}\right)\right]$$

(2.33)

The boundary condition, $A(r = a) = 0$, and radial symmetry,

$$\frac{\partial A}{\partial r}(r = a) = 0$$

lead to

$$U_1 = -1, \quad D_1 = 1 \text{ and } b_1 = 0$$
$$L_n = 0, \quad D_n = 1 \text{ and } b_n = 0$$

(2.34)

These equations form a tri-diagonal system

$$LDUA = b \tag{2.35}$$

which can be solved by double backsubstitution (Press *et al.*, 1986) and marched through time.

The age is obtained by averaging the final solution over the volume of the grain using the trapezoidal rule:

$$\text{age} = 3\int_0^1 r^2 A(r,t) dr = 3\sum_{i=1}^n w_i A_i \Delta r^3 (i-1)^2$$

$$w_i = \begin{cases} 0.5 & \text{for } i = 1, n \\ 1 & \text{otherwise} \end{cases}$$

(2.36)

2.6 Determining the diffusion parameters

The parameters entering the Arrhenius relationship (2.9), i.e. D_0 and E_a, are established by controlled-heating experiments in the laboratory. Because diffusion obeys the Arrhenius relationship, experiments at high temperatures over short times in the laboratory provide insight into diffusive behaviour at lower temperatures and correspondingly longer timescales in natural settings, as long as we can

2.6 Determining the diffusion parameters

be confident that the same diffusional processes and pathways operate in both cases.

During laboratory experiments, a sample of the mineral of interest is held at various temperatures for a fixed amount of time and the amount of gas extracted from the sample at each step is measured, from which $D(T)/a^2$ can be estimated. If the Arrhenius relation (2.9) is obeyed, then a plot of $\ln(D(T)/a^2)$ against reciprocal temperature $(1/T)$ should produce a straight line with gradient $-E_a/R$ and intercept $\ln(D_0/a^2)$ (cf. Figure 3.5). For thermochronological systems in which the diffusion domain size a corresponds to the physical grain size (e.g., for the (U–Th)/He system, cf. Section 3.2), the diffusivity D_0 of a thermochronological system can be estimated by repeating the measurements on grains with various radii. For systems in which the diffusion domain is smaller than the grain size (e.g., for most Ar–Ar systems, cf. Section 3.1), D_0/a^2 has to be estimated jointly for each sample. The constraints imposed by such experiments on the diffusion parameters and closure temperatures of the most widely used low-to-medium-temperature thermochronological systems are outlined in Chapter 3.

Because these experiments are necessarily performed at higher temperatures and on much shorter timescales than those of natural diffusion (several minutes to hours in the laboratory versus 10^6–10^8 years in Nature), their results have to be extrapolated over many orders of magnitude. Small errors in the analytical procedure may therefore result in large uncertainties in the predicted diffusion behaviour and closure temperature over geological timescales. Furthermore, there is no a-priori evidence to support the hypothesis that the same diffusion pathways and mechanisms operate on both timescales. Therefore, the results of the laboratory experiments need to be calibrated against natural settings in which thermal histories can be particularly well constrained, for instance by analysing samples from boreholes with relatively well-known temperature histories.

Tutorial 1

Use the closure temperature given in Table 1.1 to predict the (U–Th)/He apatite ages for rocks that have experienced the following cooling histories, all starting 100 Myr ago:

(i) rapid cooling from 500 to 15 °C, 40 Myr ago;
(ii) monotonic cooling from 135 to 15 °C over 100 Myr;
(iii) rapid cooling from 60 to 15 °C 20 Myr ago;
(iv) slow cooling from 100 to 60 °C over 25 Myr, isothermal conditions at 60 °C for 50 Myr, then slow cooling to 15 °C over the last 25 Myr; and
(v) slow monotonic heating from 15 to 65 °C during the first 95 Myr of the experiment, followed by a rapid cooling to 15 °C over the last 5 Myr.

Table 2.1. *Diffusion parameters to be used in Tutorial 1 to determine ages using Dodson's method and by numerical integration*

Parameter	Value
D_0/a^2	$10^{7.7}$ s^{-1}
E_a	151.46 kJ mol^{-1}
R	8.314

Use Dodson's method to determine the ages of the same rocks (the code Dodson.f is supplied on the accompanying CD). Compare the results with those obtained by solving the solid-state diffusion equation numerically (use the code MadHe.f supplied). Take the values in Table 2.1 for the diffusion parameters.

Compare the ages predicted by these three approaches. What can you deduce from this result?

3

Thermochronological systems

In this chapter, we review the three main methods used for obtaining quantitative thermochronological constraints, $^{40}Ar - ^{39}Ar$, (U–Th)/He and fission-track dating, with the aim of providing an overview of which analytical tool may be most suited for a given tectonic or geomorphic problem. The emphasis is not on an exhaustive review of all technical aspects of these methods: excellent manuals already exist for each, and we direct the reader to consult these on specific methodological questions. Instead, we seek to provide an illuminating background to the use of these techniques for the non-specialist, emphasising the accuracy and reproducibility of data, and common problems that may affect data quality.

3.1 Ar dating methods

Potassium–argon (K–Ar) dating, and the associated $^{40}Ar/^{39}Ar$ method of analysis, are among the most widely used thermochronological methods (McDougall and Harrison, 1999). Potassium is a major element that makes up about 1.5% of the Earth's crust and is abundant in common rock-forming minerals such as K-feldspars, amphiboles and micas. Potassium has three naturally occurring isotopes: ^{39}K and ^{41}K, which are stable and together make up about 99.9% of the natural abundance of K, and the radioactive isotope ^{40}K (Table 3.1).

The latter decays by the following reactions:

$$^{40}_{19}K \rightarrow {}^{40}_{18}Ar + \beta^+ \ (0.001\%)$$
$$^{40}_{19}K + \beta^- \rightarrow {}^{40}_{18}Ar \ (10.3\%) \quad (3.1)$$
$$^{40}_{19}K \rightarrow {}^{40}_{20}Ca + \beta^- \ (89.7\%)$$

Naturally occurring calcium, another abundant and major rock-forming element in the Earth's crust, is dominated by the isotope ^{40}Ca (the natural isotopic

Table 3.1. *Natural abundances of K and Ar isotopes*

Isotope	Abundance (%)
^{39}K	93.2581 ± 0.0029
^{40}K	0.01167 ± 0.00004
^{41}K	6.7302 ± 0.0029
^{40}Ar	99.60
^{38}Ar	0.063
^{36}Ar	0.337

abundance of which is 96.9%). Therefore, there are commonly large quantities of ^{40}Ca bound up in mineral structures, which are indistinguishable from the radiogenically produced ^{40}Ca. Moreover, the diffusion behaviour of Ca is complex due to its reactivity. For these reasons, the decay reaction of ^{40}K to ^{40}Ca is not generally used as a dating method, although some authors have utilised this reaction under particularly favourable conditions (low-Ca minerals, cf. McDougall and Harrison (1999)). Argon, although a secondary constituent of the Earth's atmosphere (~1% by weight), is sufficiently rare in rocks that the radiogenic component can be precisely measured. In most circumstances it can be assumed that Ar was not present in the crystal structure at its formation. Because Ar is a noble gas, it is unreactive and its diffusion behaviour is relatively simple, making it a particularly effective thermochronometer. On applying Equation (2.7) to the K–Ar system, the age equation becomes

$$t = \frac{1}{\lambda} \ln\left(\frac{^{40}\text{Ar}^*}{^{40}\text{K}} \frac{\lambda}{\lambda_e} - 1\right) \qquad (3.2)$$

where λ is the total decay constant of ^{40}K (5.543×10^{-10} yr^{-1}), λ_e is the decay constant of ^{40}K to ^{40}Ar (0.581×10^{-10} yr^{-1}) and ^{40}Ar* is radiogenically produced ^{40}Ar, that is, the amount of measured ^{40}Ar corrected for any potentially extraneous component. The abundances of both ^{40}K and ^{40}Ar* thus have to be measured in order to obtain an age.

In the classical K–Ar approach to geochronology, the concentration of K in a sample is measured by wet chemical methods or flame photometry, from which the abundance of ^{40}K in a measured aliquot can be deduced through the constant abundance ratios of K isotopes in nature (Table 3.1). ^{40}Ar is measured by fusion of a separate aliquot of the sample, purification of the noble gases released, and measurement of their isotopic composition in a mass spectrometer. The measurement of isotopic abundance is calibrated by adding a spike (i.e. a known amount) of ^{38}Ar to the Ar from the sample and measuring the ratio of

3.1 Ar dating methods

^{40}Ar to ^{38}Ar. The presence of trace amounts of atmospheric gas, of which Ar is the third most abundant component, adhering to the sample as well as interior surfaces of the extraction system and mass spectrometer, can contribute significantly to the amount of ^{40}Ar measured. A correction to the measured ^{40}Ar abundance must therefore be applied to account for this possible contamination. This is performed through measurement of the stable, non-radiogenic isotope ^{36}Ar relative to the ^{40}Ar released. The ratio of these two isotopes is constant in the atmosphere (^{40}Ar/^{36}Ar = 295.5%; cf. Table 3.1). Assuming that all ^{36}Ar present is of atmospheric origin, the measured ^{40}Ar extracted can be corrected for this contamination through the relationship:

$$^{40}\text{Ar}^* = {}^{40}\text{Ar}_{measured} - 295.5 \times {}^{36}\text{Ar}_{measured} \tag{3.3}$$

If the system has cooled to temperatures such that diffusive loss of ^{40}Ar* is slower than the increase in ^{40}Ar* caused by decay of ^{40}K, a component of the ^{40}Ar* formed in the sample is retained over geological timescales, and a finite 'age' will begin to accumulate in the sample (cf. Section 2.2). The significance of ages derived from K–Ar measurements depends strongly on the relative speed of this cooling and the consequent duration of the transition from open-system to closed-system behaviour (see below).

Diffusive transport alters the distribution of ^{40}Ar* within crystals and leads to its loss from the system. ^{40}Ar* distributions within a grain population can also be modified by alteration or recrystallisation, and the incorporation of ^{40}Ar-rich fluid inclusions or back-diffusion into a crystal lattice due to high external partial Ar pressures can lead to the incorporation of 'excess' argon (that is, ^{40}Ar from a source other than *in situ* decay of ^{40}K within a sample, inheritance, or atmospheric contamination). Neither the extent, nor indeed the occurrence, of these forms of altered argon distribution can be directly assayed by the bulk extraction techniques adopted in K–Ar dating, restricting the quantitative application of this approach to the limited case of samples for which a simple thermal and petrological history can be assumed.

These limitations are directly addressed by the ^{40}Ar – ^{39}Ar analytical technique. As discussed below, the possibility of progressively extracting and analysing argon with this method provides the potential to resolve the spatial distribution of argon in a sample and, with it, the corresponding thermal and geological histories of samples.

The ^{40}Ar – ^{39}Ar technique has the advantage that no knowledge of the absolute K and Ar concentrations is required. Instead, the amount of ^{39}K is determined by a proxy, in the form of ^{39}Ar produced by irradiating the sample with fast neutrons in a nuclear reactor. ^{39}Ar is produced through the reaction

$$^{39}_{19}\text{K} + {}^{1}_{0}\text{n} \rightarrow {}^{39}_{18}\text{Ar} + {}^{1}_{1}\text{p} + Q \tag{3.4}$$

where Q represents the energy released. After irradiation, the argon content of a sample is extracted by progressive heating under vacuum in a furnace or laser cell, and, after correction for various potential isotopic interferences, the ratios of ^{40}Ar to ^{39}Ar released during each step of the experiment are compared to derive an age. The ^{40}Ar/^{39}Ar ratio is determined by

$$\frac{^{40}\text{Ar}}{^{39}\text{Ar}} = \frac{\lambda_e}{\lambda} \frac{^{40}\text{K}}{^{39}\text{K}} \frac{e^{\lambda t} - 1}{\alpha} \tag{3.5}$$

where α is a proportionality factor that depends on the neutron dose and the capture cross-section of ^{39}K for fast neutrons. Because α is constant for each irradiation, and λ_e/λ and ^{40}K/^{39}K are also constants, they can be combined into a single parameter J (known as the irradiation factor), which can be defined as

$$J = \frac{\alpha}{(\lambda_e/\lambda)(^{40}\text{K}/^{39}\text{K})} \tag{3.6}$$

A value for J can be estimated by irradiating a mineral standard of known age together with the unknown samples. In this way, no exact knowledge of the received neutron fluence is required, since this can be assayed in comparative terms by employing the relative gas evolution from the unknown sample and the well-constrained age standard. The age equation then becomes

$$t = \frac{1}{\lambda} \ln\left(1 + J \frac{^{40}\text{Ar}}{^{39}\text{Ar}}\right) \tag{3.7}$$

Ar can be extracted stepwise by heating the sample over a range of temperatures below that of fusion, which presents a powerful advantage over bulk extraction and analysis in that it provides insight into the relative spatial distribution of ^{40}Ar. ^{39}Ar and ^{40}Ar behave essentially identically during extraction; any gas released during a given heating step thus contains both ^{40}Ar and a direct proxy for the ^{40}K content of the correlative area of the sample. Any departure from uniformity in the ratio of these components during the various heating stages of the experiment can therefore be read as a modification of the ^{40}Ar distribution expected from the decay of ^{40}K alone. By investigating sample behaviour over a range of temperatures in the laboratory, a pattern of argon release is derived that can be interpreted geologically (Figure 3.1); either as partial retention of Ar (lower ages for low-temperature steps) or as the incorporation of excess Ar (higher ages for low-temperature steps).

Alternatively, the spatial distribution can be mapped directly: Ar can be extracted from the sample by a spot-fusion technique using a laser coupled to the extraction line and mass spectrometer, which allows determination of the ^{40}Ar*/^{39}Ar ratio in very small samples (Figure 3.2). This makes it possible to date multiple phases of deformation by spot-dating deformation tails on metamorphic

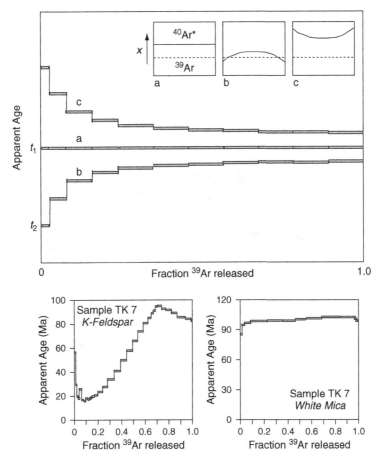

Fig. 3.1. ^{40}Ar/^{39}Ar age spectra obtained by step-heating experiments. The upper panel shows the three principal types of spectra that may be obtained: (a) an ideal flat age spectrum indicating rapid cooling at time t_1; (b) a monotonically rising age spectrum indicating argon loss and either partial resetting at time t_2 or slow cooling from t_1 to t_2; and (c) uptake of extraneous ('excess') argon leads to progressively decreasing apparent ages. Insets show schematic distributions of ^{40}Ar* and ^{39}Ar in the sample for these three cases. Lower panels show age spectra from a sample from the Hohonu Range, South Island, New Zealand. Left: the K-feldspar spectrum indicates uptake of excess Ar (high ages for the first few extraction steps) followed by a staircase-like spectrum with ages increasing from ∼16 Myr for low-temperature steps to ∼90 Myr at fusion. Right: the white-mica spectrum for the same sample is a relatively flat spectrum, with initial ages of ∼85 Myr over the first three heating steps, rising to a plateau at ∼100 Myr for the remainder of the extraction steps. These data have been interpreted as indicating rapid cooling from ≥350 °C down to ∼250 °C at 100–90 Myr (closure of white mica and beginning of Ar retention in K-feldspar) followed by a second rapid cooling phase from ∼15–20 Myr ago onwards. Modified from Batt et al. (2004); data from Reiners et al. (2004).

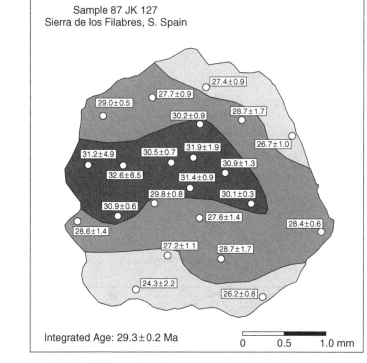
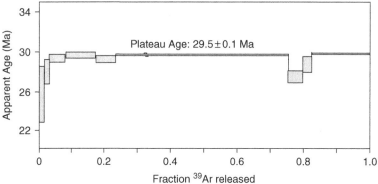

Fig. 3.2. ^{40}Ar*/^{39}Ar laser spot-fusion analysis of a single phengite grain from the Sierra de los Filabres, southern Spain. The upper panel shows an image of a grain with dated fusion spots and interpreted age contours: ages decrease from ≥ 30 Myr in the grain core to ~ 25 Myr in the rim. The lower panel shows stepheating analysis of the other half of the grain for comparison: ages increase from ~ 25 Myr for the first few extraction steps to a well-defined plateau at ~ 30 Myr for the remainder of the extraction steps. These data have been interpreted as indicating rapid cooling below ~ 350 °C at 30 Myr followed by minor thermal resetting at ~ 25 Myr. Modified from de Jong *et al.* (1992). Reproduced with permission from Elsevier.

minerals and to construct $P-T-t$ paths by dating mineral pairs in local equilibrium that provide precise $P-T$ information (Müller, 2003).

Diffusion behaviour of Ar

Empirical constraints on the diffusivity of Ar in geological materials are comprehensively reviewed by McDougall and Harrison (1999). Both the volume of published studies and the perceived level of understanding of this process are greatest for alkali feldspar. This mineral structure is stable during heating under vacuum at temperatures below melting (e.g., Fitzgerald and Harrison, 1993; Lovera *et al.*, 1993). Most of the Ar trapped within K-feldspar can thus be driven off in the lab via existing diffusion pathways, allowing experimental determination of diffusion characteristics for each individual sample during the routine procedure of step-heating analysis (Lovera *et al.*, 1991; Richter *et al.*, 1991).

Results of such argon-diffusion studies (e.g., Lovera *et al.*, 1997, 2002) have demonstrated that most K-feldspars possess diffusion properties that vary as a function of temperature. Although yielding initially linear Arrhenius relationships, indicating argon-diffusion properties similar to those obtained for gem-quality orthoclase by Foland (1974), this simple behaviour breaks down at higher temperatures for most samples. The most successful explanation of this behaviour to date is the multiple-diffusion-domain (MDD) model of Lovera *et al.* (1989). The MDD model proposes the presence of a distribution of discrete non-interacting domains of varying length scales in the K-feldspar crystal lattice. This model is able to relate the degassing systematics of reactor-produced ^{39}Ar observed during laboratory heating to the age spectrum of a sample (reflecting loss and retention of ^{40}Ar on geological timescales), providing a basis for the reconstruction of thermal histories over a broad temperature range ($\sim$150–350 °C) applicable to the middle crust (Lovera *et al.*, 1989, 1997).

The experimental investigation of argon retentivity in other minerals is complicated by the hydrous (e.g., micas, amphiboles) and/or iron-bearing (e.g., biotite, hornblende) character of many K-bearing mineral phases. Rather than out-gassing by volume-diffusion processes, such structures experience structural breakdown by dehydration and delamination reactions, producing corresponding argon release (e.g., Vedder and Wilkins, 1969), when heated in a vacuum. This lack of stability prevents effective constraint of their natural diffusive properties by simple step-heating experiments (Brandt and Voronovskiy, 1967).

Overcoming this instability requires heating the sample under hydrothermal conditions to preserve the integrity of the mineral structure on the timescale required for measurable diffusive transfer of argon to occur – which can be up to months at the relevant temperatures in the laboratory (McDougall and Harrison,

Table 3.2. *Ar-diffusion parameters*

Mineral	D_0 (cm^2 s^{-1})	E_a (kJ mol^{-1})	Reference
Phlogopite (Ann$_4$)	$0.75^{+1.7}_{-0.52}$	242 ± 11	Giletti (1974)
Biotite (Ann$_{56}$)	$0.077^{+0.21}_{-0.06}$	196 ± 9	Harrison et al. (1985)
Biotite (Ann$_{56}$)	$0.015^{+0.022}_{-0.005}$	188 ± 12	Grove and Harrison (1996)
Biotite (Ann$_{56}$)	$0.075^{+0.049}_{-0.021}$	197 ± 6	Combined data of Harrison et al. (1985) and Grove and Harrison (1996), presented in McDougall and Harrison (1999)
Biotite (Ann$_{71}$)	$0.40^{+0.96}_{-0.28}$	211 ± 9	Grove and Harrison (1996)
Muscovite	$0.033^{+0.213}_{-0.029}$	183 ± 38	Hames and Bowring (1994)
Hornblende	$0.06^{+0.4}_{-0.01}$	276 ± 17	Harrison (1981)

1999). By comparing the amount of ^{40}Ar* remaining in the sample after the period of experimental heating with the concentration in the starting material, the fractional loss of ^{40}Ar can be determined, allowing calculation of a model diffusion coefficient for the sample.

Suitable hydrothermally buffered determinations of argon-diffusion characteristics have been carried out for phlogopite by Giletti (1974) and for biotite by Harrison et al. (1985) and Grove and Harrison (1996), both of which phases are commonly dated by the Ar method (see Table 3.2). Results of studies on grain fractions of varying size indicate that the critical length scale for diffusion of argon in biotite-series minerals is $\sim 500\,\mu$m. Physical grain size defines the argon-diffusion volume for crystals smaller than this (Hess et al., 1993), but the $500-\mu$m length scale remains the limiting dimension for larger grains. Since the retention and loss of argon are thus often controlled at a sub-grain level, the pertinent geometrical relationship to use in evaluating diffusive loss is not straightforward. Rather than direct evaluation, this property is usually derived empirically from numerical modelling of the data, and is often one of the more debated aspects of argon-diffusion studies.

Diffusion of argon in biotite is also likely to be significantly influenced by variations in chemical composition. Different proportions of the ions that can be present in the biotite structure (Fe^{3+}, Fe^{2+}, Mg^{2+}, Al^{3+}, F^-, Cl^- and OH^-) cause up to 5% variation in the molar volume of the crystal (Hewitt and Wones, 1975; Grove and Harrison, 1996). Using an ionic porosity model of argon diffusion tested against the available experimental evidence, Grove and Harrison (1996) predicted that this range of biotite structure may cause argon-diffusion coefficients to vary by an order of magnitude.

Biotite is often ascribed a nominal closure temperature for argon of ∼300 °C (Hodges, 1991), but Grove and Harrison (1996) suggest that the predicted compositional variation can allow this parameter to range as high as 450 °C. Although attempts to use natural settings to constrain diffusion properties in biotite have been made (Hurley et al., 1962; Hart, 1964; Westcott, 1966; Hanson and Gast, 1967), these lack sufficient precision to test adequately the experimental diffusion data and predictions of their thermal significance, due to large uncertainties in various aspects of the thermal histories of the regions used (McDougall and Harrison, 1999).

Although argon release from muscovite has been examined under hydrothermal conditions (Hart, 1964; Brandt and Voronovskiy, 1967), the most detailed hydrothermal diffusion study undertaken for this mineral is reported in a widely cited, but unpublished, Brown University thesis (Robbins, 1972). Various authors have re-examined the Robbins (1972) data, and recast interpretations of it using different geometric models of diffusive loss (Hames and Bowring, 1994; Lister and Baldwin, 1996).

Uncertainties in these experimentally derived diffusion properties prevent *ab initio* calculation of meaningful closure temperatures for muscovite (Hames and Bowring, 1994; Lister and Baldwin, 1996). Empirical results generally show muscovite to have Ar ages concordant with or older than those of coexisting biotite, and it has become common practice to assume a nominal closure temperature for muscovite ∼50 °C higher than that of biotite in the same sample (Purdy and Jäger, 1976; Cliff, 1985; Hodges et al., 1996). McDougall and Harrison (1999), in their review of this subject, sound a note of caution on this practice, in the light of the demonstrable variation in effective biotite closure temperature by ∼150 °C under differing conditions.

Hydrothermal studies of argon diffusion in hornblende have been performed by Harrison (1981) and Baldwin et al. (1990), confirming results of field-based studies and vacuum extraction experiments suggesting that this is the most argon-retentive of the phases commonly dated by the K–Ar method (see Table 3.2), with a closure temperature ranging from ∼450 to 600 °C depending on the geological cooling rates experienced. Modelling also suggests the occurrence of significant changes in ionic porosity and corresponding argon-diffusion properties with compositional variation, which are dominated by the effects of Mg content and A-site occupancy (Dahl, 1996).

Modelling routines

We provide two subroutines (Biotite.f90 and Muscovite.f90) to compute biotite and muscovite K–Ar ages based on Dodson's approximation (i.e., Equation (2.21)).

More sophisticated models have been developed and coded for this purpose by other authors, such as the MacArgon program which is available by downloading from http://www.earth.monash.edu.au/macargon/ and described in Lister and Baldwin (1996), which computes Ar ages and spectra for biotite, muscovite and hornblende from a given thermal history. Another useful program is Arvert, which was developed by P. Zeitler and is available by downloading from http://www.ees.lehigh.edu/EESdocs/geochronology.shtml This program inverts K-feldspar Ar ages and spectra for thermal history, based on the multi-domain method described in Lovera et al. (1989).

3.2 (U–Th)/He thermochronology

The production of ^{4}He (α particles) from uranium (U) and thorium (Th) series decay in rocks and minerals was the first geochronological dating method to be proposed early in the twentieth century (Rutherford, 1907; Soddy, 1911–1914). However, at that time geoscientists were pursuing only 'absolute' or formation ages of rocks, notably in a quest to constrain the age of the Earth (Holmes, 1913). Since He diffuses easily out of the mineral lattice (a phenomenon that was not understood at the time), ages determined using U, Th and He measurements were consistently much younger than those calculated using the U–Pb couple. The use of He as a geochronometer consequently came to be considered unreliable and was abandoned (Hurley et al., 1956).

Interest in the technique has been revived since Zeitler et al. (1987) proposed that the diffusive loss of He could be quantified and that He ages could be used to constrain cooling through very low temperatures. Subsequent diffusion experiments (Wolf et al., 1996; Farley, 2000) have demonstrated that the apatite (U–Th)/He thermochronometer is sensitive to temperatures as low as 40 °C, with effective closure occurring around 70 °C, depending on the cooling rate and mineral grain size. Comparisons of apatite (U–Th)/He and fission-track ages (Warnock et al., 1997; House et al., 1999; Stockli et al., 2000) have confirmed this range of relative temperature sensitivity on geological timescales (see Figure 3.10 later). More recent experiments to determine the thermal response of other accessory minerals have shown that the effective closure temperature is ~160–200 °C in zircon (Reiners et al., 2002, 2004) and ~190–220 °C in titanite (Reiners and Farley, 1999) (see Table 1.1). The low closure temperatures of the (U–Th)/He system, especially in apatite, make it particularly sensitive to near-surface cooling and thermal perturbations. This increased resolving power towards the low end of the temperature spectrum able to be constrained by thermochronology has gained the method considerable interest from the Earth-science community over the last few years, especially from those in the field of geomorphology.

3.2 (U–Th)/He thermochronology

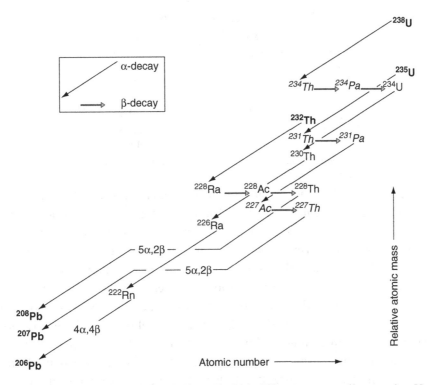

Fig. 3.3. Decay chains of U and Th to Pb. In boldface are naturally occuring U, Th and Pb isotopes at the beginning and end of the chain; in italics are isotopes with half-lives so short that they cannot normally be measured.

He is produced (as α particles) in a cascade of reactions from ^{238}U, ^{235}U and ^{232}Th (Figure 3.3). The He ingrowth equation is

$$^4\text{He} = 8 \times {}^{238}\text{U}(e^{\lambda_{238}t} - 1)$$
$$+ 7 \times \frac{^{238}\text{U}}{137.88}(e^{\lambda_{235}t} - 1)$$
$$+ 6 \times {}^{232}\text{Th}(e^{\lambda_{232}t} - 1) \qquad (3.8)$$

where ^{4}He, ^{238}U and ^{232}Th are the measured present-day abundances of these isotopes, t is the He age and the λs are the decay constants. The constants preceding the U and Th abundances account for the multiple α particles emitted for each decay series and the factor 137.88 is the present-day ^{238}U/^{235}U ratio. Natural abundances and decay constants of the parent isotopes are given in Table 3.3.

In order to obtain a He age of a sample, the present-day abundances of U, Th and He must therefore be measured. He is extracted from the sample by heating

Table 3.3. *Natural abundances and decay constants of U and Th isotopes*

Isotope	Abundance (%)	λ (yr^{-1})
^{238}U	99.2745	1.55×10^{-10}
^{235}U	0.7200	9.85×10^{-10}
^{234}U	0.0055	2.83×10^{-6}
^{232}Th	100	4.95×10^{-11}

under vacuum in a furnace or laser cell at temperatures well below those of grain melting or physical breakdown. The evolved gas is cleaned and purified, and the He content measured on a noble-gas mass spectrometer. As for the analysis of Ar in the Ar-based methods, measurement of the ^{4}He extracted is performed relative to a spike comprising in this case a known abundance of ^{3}He. After He-extraction, the samples are recovered from the extraction line and dissolved. U and Th in the resulting solution are then measured by inductively coupled plasma mass spectrometry. This approach allows U, Th and He to be measured on the same aliquot, eliminating the issue of inhomogeneity between grains and meaning that precise grain masses need not be measured. With the precision attainable using currently available facilities, single grains of zircon can be routinely dated. For apatite and titanite, which are characterised by much lower U and Th concentrations than those in zircon, multi-grain aliquots are generally required, except where sufficient time has elapsed since closure (typically several tens of millions of years) for sufficient helium ingrowth to occur.

Because of the high diffusivity of He through most minerals and the relative rarity of He in the environment (with a concentration in the atmosphere of only 5 p.p.m.), initial or excess He is usually not a concern. However, two factors are known to bias the measured He age strongly: α-ejection and (U–Th)-rich inclusions. The first is related to the fact that the α particles produced by nuclear decay have a kinetic energy that allows them to travel up to 20 μm through the crystal lattice (Farley, 2002), and potentially to be ejected from the mineral (Figure 3.4). This effect is corrected for by using a numerical model of α-ejection. Farley *et al.* (1996) proposed an approximate analytical model of α-ejection, for simplified grain geometries and homogeneous U distributions. More recently, Meesters and Dunai (2002a, 2002b) have proposed a finite-difference solution for realistic (finite) grain geometries and inhomogeneous U distributions. Whatever method is chosen, correcting for α-ejection requires monitoring the sizes (and preferably also shapes) of the minerals analysed.

The second potential problem is related to the fact that many apatite grains contain minute inclusions of actinide-rich minerals such as zircon and monazite

3.2 (U–Th)/He thermochronology

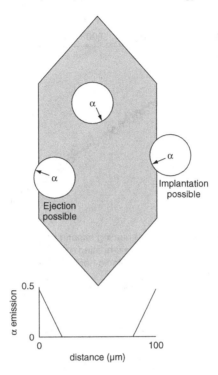

Fig. 3.4. A schematic representation of α-ejection in apatite. Of the three α particles shown, one will be retained in the crystal, one can be ejected, and one can recoil into the crystal from its surroundings. The latter situation is rare due to the low concentrations of U and Th in phases other than the minerals used for dating. The lower plot shows the predicted probability of α-ejection along a profile across the crystal. Modified from Farley (2002). Reproduced with permission by the Mineralogical Society of America.

that contribute to the ^{4}He abundance in the sample but, because they are not fully dissolved by the standard chemical dissolution methods used on apatite, do not contribute to the U and Th measurements. Samples that contain such inclusions will therefore present 'parentless' He and yield excessively old ages. Minimising this problem requires very careful sample selection by screening every grain to be dated for even the smallest inclusions under a high-magnification optical microscope.

Diffusion behaviour of He

Step-heating experiments have been performed to constrain the diffusivity of helium in apatite (Farley, 2000), titanite (Reiners and Farley, 1999) and zircon (Reiners *et al.*, 2002). For apatite there is a good linear correlation of $\ln(D(T)/a^2)$ with $(1/T)$ for temperatures below ~ 600 K (Figure 3.5), indicating

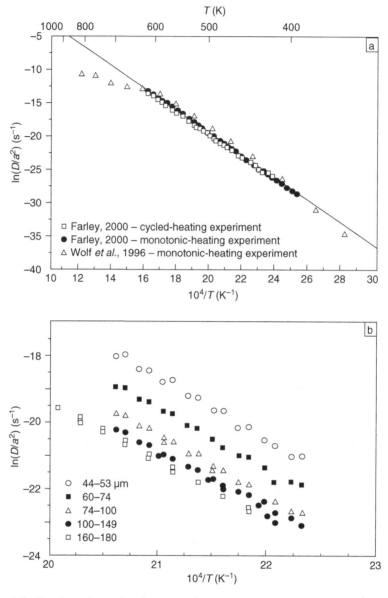

Fig. 3.5. Results of step-heating experiments on He extraction from apatite: (a) Measured $D(T)/a^2$ as a function of reciprocal temperature. The data show the occurrence of good Arrhenius behaviour for temperatures ≤ 600 K. (b) The dependence on grain size: results for aliquots of various grain sizes indicated by the key. After Farley (2000). Reproduced with permission from the American Geophysical Union.

Table 3.4. *Estimates of diffusion parameters for the (U–Th)/He system in various minerals*

Mineral	$D_0 (m^2 s^{-1})$	$E_a (kJ\ mol^{-1})$	Reference
Apatite	$5^{+4}_{-2} \times 10^{-3}$	138 ± 2	Farley (2000)
Titanite	$\sim 6 \times 10^{-3}$	186 ± 14	Reiners and Farley (1999)
Zircon	$4.6^{+8.7}_{-3.0} \times 10^{-5}$	168 ± 5	Reiners et al. (2004)

simple Arrhenius behaviour. Above this temperature, both the activation energy and D_0/a^2 are lower than expected from extrapolating the low-temperature data. Farley (2000) suggested that this may be a result of defect annealing in apatite at these temperatures, since fission tracks in apatite also anneal rapidly over this temperature range in the laboratory. The measured D_0/a^2 scales well with the square of grain size, suggesting that the diffusion domain for He in apatite is the whole grain. Estimated diffusion parameters for apatite are given in Table 3.4.

Extrapolation of these parameters suggests that the partial-retention zone for He in apatite lies between ~ 40 and $70\,°C$ (Farley, 2000). Results of several (U–Th)/He studies in boreholes (Warnock et al., 1997; House et al., 1999) have supported this temperature range in natural settings, within the uncertainty of the thermal histories experienced by the borehole samples. The position of the He partial-retention zone has been compared with the partial-annealing zone of fission tracks in apatite and it was found that their relative positions were consistent with the laboratory-derived estimates (House et al., 1997; Stockli et al., 2000)(see Figure 3.10 later).

Much less experimental data is available for titanite and zircon. Reiners and Farley (1999) performed diffusion experiments in titanite and found relatively simple Arrhenius behaviour, comparable to that in apatite. On this basis, they suggest that the diffusion domain in titanite is also the whole grain. In zircon, the diffusion behaviour is more complicated. Cycled temperature experiments (that is, experiments in which the sample is first heated to some peak temperature and then subjected to a series of heating steps with progressively decreasing peak temperatures) showed that there are lower diffusivities during the 'cooling' steps (that is, steps in which the peak temperature is reduced relative to the previous step) than during the heating steps (Reiners et al., 2002). This effect could be related either to the operation of diffusion on multiple length scales or to an effect of α-damage. Reiners et al. (2004) preferred the former mechanism and estimated the diffusion parameters for zircon. A comparison of zircon (U–Th)/He ages with thermal histories derived from $^{40}Ar-^{39}Ar$ measurements (Reiners et al., 2004) and a comparison of the positions of the He partial-retention zones for titanite and zircon with apatite and zircon partial fission-track annealing zones (Reiners *et al.*, 2002) suggest that the closure temperature for He both in titanite and in zircon is around 200 °C, which is consistent with the diffusion parameters found in the experiments.

Mad_He.f90

The algorithm described in Section 2.5 has been incorporated into the subroutine Mad_He.f90, which models the integrated production and diffusive loss of helium from an apatite sample to produce a synthetic (U–Th)/He age. Values for the diffusion parameters D_0/a^2 and activation energy, E, can be set by the user, and the input for the subroutine comprises a thermal history in the form of two arrays of length *ntime*, the first recording the temperature at each step, the second the corresponding time values.

Mad_He.f90 provides a finite-difference solution to the coupled He-ingrowth and diffusion equations for homogeneous U and Th distributions and simplified (spherical) grain geometries, which leads to a sufficiently accurate age prediction for most situations. Alternative approaches exist: Wolf *et al.* (1998) provided an approximate analytical solution to the problem that is very simple to code. For more complex situations, the code DECOMP (Dunai *et al.*, 2003) provides a general solution to the problem based on the approach taken by Meesters and Dunai (2002a, 2002b), which handles non-uniform U and Th distributions, realistic grain geometries and α-ejection.

3.3 Fission-track thermochronology

Uranium decays not only by α and β emission; a small proportion of ^{238}U decays by splitting of the atom, or fission. Upon fission, two positively charged high-energy nuclei are created that are propelled away from each other, creating a single linear trail of ionisation damage referred to as a fission track. Fission-track thermochronology uses these damage zones, which can be revealed by chemical etching and counted under an optical microscope, as a daughter product: each track represents a fission-decay event. Over the past 20 years, fission-track thermochronology has become established as a widely used technique for constraining the low-temperature thermal histories of rocks. Excellent reviews of the technique have been provided by Brown *et al.* (1994), Gallagher *et al.* (1998), Gleadow and Brown (2000), Hurford (1991) and Ravenhurst and Donelick (1992), while in-depth discussions of the theory can be found in Fleischer *et al.* (1975) and Wagner and Van den Haute (1992).

3.3 Fission-track thermochronology

Fission tracks are originally 10–20 μm long and only 25–50 Å wide, their length depending on the density of the crystal lattice (i.e. ~11 μm in zircon; ~16 μm in apatite). Owing to their narrow width, fission tracks in their natural state ('latent' tracks) are visible only using transmission electron microscopy (Paul and Fitzgerald, 1992). They can be 'revealed', that is, treated to become visible under an optical microscope with $\geq 1000\times$ magnification, by polishing and chemically etching an internal surface of the crystal (Figure 3.6). Since each spontaneous fission event creates one fission track, the track density is a function of the rate of fission decay, the concentration of ^{238}U and the fission-track age of the sample. The fission-track age equation can be written

$$t = \frac{1}{\lambda_D} \ln\left(\frac{\lambda_D}{\lambda_f} \frac{N_s}{^{238}U} + 1\right) \tag{3.9}$$

where λ_D is the total decay constant for ^{238}U, λ_f is the spontaneous-fission decay constant for ^{238}U, N_s is the number of spontaneous-fission tracks present in the sample and ^{238}U is the number of ^{238}U atoms in the sample. The half-life for spontaneous fission of ^{238}U lies between 8.5×10^{15} and 9.9×10^{15} yr (Fleischer et al., 1975; Wagner and Van den Haute, 1992), which is more than six orders of magnitude higher than that for α-decay of ^{238}U (4.5×10^9 yr; Table 3.3). The total decay constant for ^{238}U is therefore very close to that for α-decay.

A simple and accurate way to determine the uranium concentration is to irradiate the sample with thermal neutrons in a nuclear reactor. This causes a proportion of the isotope ^{235}U present to undergo fission, inducing a new set of tracks. The ^{235}U/^{238}U isotope ratio is constant (Table 3.3), so the ^{238}U abundance can be

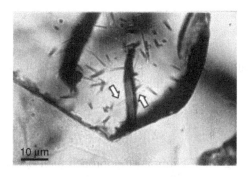

Fig. 3.6. A photomicrograph of a polished and etched section through an apatite crystal. Etched fission tracks intersecting the surface are clearly visible as cones. Arrows point to a horizontal confined track.

calculated from the induced track density. The fission-track age equation then becomes

$$t = \frac{1}{\lambda_D} \ln\left(1 + \frac{\lambda_D \phi \sigma c I \rho_s}{\lambda_f \rho_i}\right) \qquad (3.10)$$

in which c is a geometry factor ($c = 2\pi/(4\pi)$ for the external-detector method, see below); $I = {}^{235}\text{U}/{}^{238}\text{U}$ is the isotope abundance ratio (7.253×10^{-3}), σ is the thermal-neutron fission cross-section for ${}^{235}\text{U}$ ($580.2 \times 10^{-24}\,\text{cm}^{-2}$), ϕ is the thermal-neutron fluence (in $\text{cm}^{-2}\,\text{s}^{-1}$), ρ_s is the spontaneous track density and ρ_i is the induced track density in the sample (Hurford, 1990). The thermal-neutron fluence received by the sample during irradiation is constrained by irradiating a dosimeter glass with known U content at the same time as the sample and determining the track density in the dosimeter, ρ_d. The relationship between neutron fluence and neutron-induced track density in the dosimeter is linear: $\phi = B \times \rho_d$, where B is a proportionality constant.

The most commonly used fission-track dating technique is the external-detector method (Hurford and Green, 1982; Hurford, 1990), in which the sample is first polished and etched to reveal spontaneous tracks, after which an external detector (usually a thin sheet of low-U mica) is attached to the sample and the package is sent off for irradiation. After irradiation, the detector is etched to reveal the induced tracks, after which a mount is made that includes the sample and its image revealed in the external detector (Figure 3.7). This approach has the advantage that inter-grain variability in track densities is revealed, so statistical tests of the significance of the fission-track age can be performed.

There are parameters in the fission-track age equation (3.10) that are either poorly known (λ_f) or difficult to determine accurately (ϕ). This problem is circumvented if a ζ-calibration approach (Hurford and Green, 1983) is adopted. The constants in the equation are then taken together in a ζ factor:

$$\zeta = \frac{\phi \sigma I}{\lambda_f \rho_d} \qquad (3.11)$$

ζ is calibrated against accepted standard samples of a known and well-characterised age t_{std} that are irradiated with the sample (Hurford, 1990):

$$\zeta = \frac{e^{\lambda_D t_{\text{std}}} - 1}{\lambda_D (\rho_s/\rho_i)_{\text{std}} c \rho_d} \qquad (3.12)$$

The age equation using the ζ-calibration approach then becomes

$$t = \frac{1}{\lambda_D} \ln\left(1 + \frac{\lambda_D \zeta \rho_s c \rho_d}{\rho_i}\right) \qquad (3.13)$$

3.3 Fission-track thermochronology

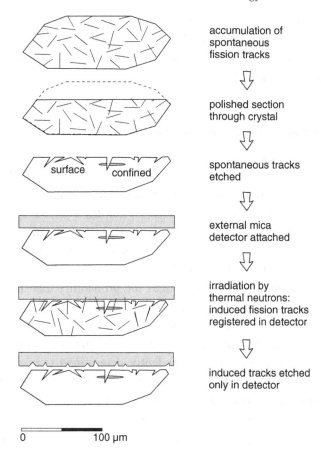

Fig. 3.7. A schematic representation of the external-detector method. After Gallagher et al. (1998). Reproduced with permission from Annual Reviews.

In the external-detector method, an age is calculated for each grain analysed (usually about 20 grains in basement samples). These can then be combined to give a sample age in different ways: the *mean* age is simply the mean of the individual grain ages, whereas the *pooled* age is calculated by pooling the spontaneous and induced fission-track counts in the individual grains (Green, 1981). The latter is meaningful only if the individual grain ages form a single age population, which is not necesarily the case: single-grain ages can be dispersed in sediments if the grains come from different sources and record different cooling histories, or in slowly cooled samples where grains with slightly different chemical compositions have been annealed to different degrees (see the next section). To test whether the single-grain ages form a unique population, a statistical χ^2-test is usually performed (Green, 1981). The most widely used approach in recent studies is to calculate a *central* age from the logarithmic mean of the single-grain ages and the associated age dispersion, which is the relative standard error of the single-grain ages (Galbraith and Laslett, 1993). The generally quoted (1σ or 2σ) error on the sample age is propagated from the number of tracks counted in the sample, the external detector and the dosimeter.

Annealing of fission tracks and confined track-length distributions

The damage zone comprising a fission track in the crystal lattice is not stable and will tend to be repaired. This occurs by a diffusive process called annealing, during which atoms and electrons move through the crystal lattice towards the ionised track. As for all diffusive processes, fission-track annealing takes place at strongly temperature-dependent rates. As a result of annealing, the etchable length of a track, which is initially similar for all tracks in a given mineral structure, will be progressively shortened (Green et al., 1986; Carlson, 1990). Because the mean length of the tracks in a sample determines the probability that they intersect an internal surface, the track density (and thus the apparent fission-track age) is also reduced during annealing (Green, 1988).

Annealing under geological conditions has been studied in apatites recovered from drill-cores in basins with independently determined temperature histories (Gleadow and Duddy, 1981; Naeser, 1979, 1981). These studies have shown that the temperature of total annealing is $120 \pm 10\,°C$ for apatite (Figure 3.8). Above these temperatures, annealing takes place at a faster rate than track production, so the effective apatite fission-track age remains perpetually zero. Fission-track annealing temperatures for zircon and titanite are significantly higher but are known with much less certainty. This lack of empirical constraint arises because the annealing temperatures for zircon and titanite are encountered only in ultra-deep drillholes and hence the relevant data are much sparser than for apatite.

Figure 3.8 shows how the fission-track age and mean track length decrease with increasing temperature under thermally stable conditions. The amount of annealing increases non-linearly with temperature (Figure 3.8). Below $\sim 60\,°C$ annealing rates are very slow but they increase rapidly at depths corresponding to the temperature range of $60-120\,°C$, an interval that has been termed the partial-annealing zone (PAZ) (Naeser, 1979; Wagner, 1979) and that is analogous to the partial-retention zone in isotopic systems. The rates of annealing in apatite samples are influenced by their chemistry: the $Cl/(F+Cl)$ ratio dominates this relationship (Green et al., 1986, 1989b), although the substitution of other anions (OH) and cations (rare-earth elements, Mn, Sr) also plays a role (Carlson et al., 1999; Barbarand et al., 2003). Moreover, the annealing rate also appears to depend on the crystallographic orientation of the tracks, with tracks orthogonal to the C-axis of the mineral annealing more rapidly than those parallel to the

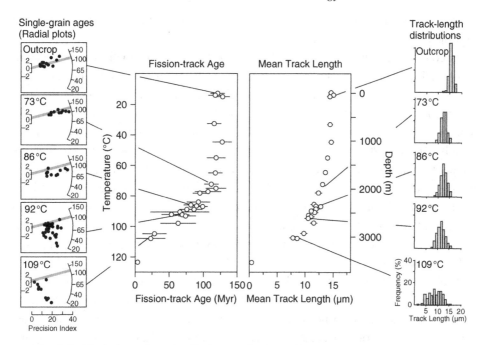

Fig. 3.8. Variation of apatite fission-track age and mean track length with temperature downhole for samples from the Otway Basin, southeast Australia. The stratigraphic age for these samples is 120 Myr. Also shown are histograms of track-length distributions, as well as radial plots that show the dispersion from the mean of the individual grain ages as a function of their relative precision (Galbraith, 1990). Grey bands on radial plots indicate depositional age. Note the wider track-length distributions and larger spread in single-grain ages with increasing annealing. Modified from Gallagher et al. (1998); data from Gleadow and Duddy (1981), Green et al. (1989a) and Brown et al. (1994). Reproduced with permission from Annual Reviews.

C-axis (Green et al., 1986; Donelick et al., 1999). The mean width of fission-track etch pits has been proposed as a proxy measure for the varying annealing kinetics arising from compositional and structural variation (Ketcham et al., 1999; Barbarand et al., 2003).

Because of the chemical and crystallographic dependences of annealing rates, the track-length distributions (measured as the standard deviation σ) become wider and single-grain ages become more dispersed with increasing annealing, as shown in Figure 3.8.

Recently, the possible effect of pressure on fission-track annealing has been the subject of controversy. On the basis of results from a set of laboratory experiments, Wendt et al. (2002) have argued that higher pressures may significantly increase track stability. Their study has been criticised (Kohn et al., 2003) for its experimental design and for lacking consideration of earlier work, but the argument has not been settled.

Zircons have a much less variable chemistry than do apatites and no compositional influence on annealing rates has been observed. However, estimates of total annealing temperatures for zircon based on calibration with other thermochronometers (Hurford, 1991) as well as some natural data (Zaun and Wagner, 1985; Tagami *et al.*, 1995; Tagami and Shimada, 1996) have varied widely, from ~ 200 to $\geq 280\,°C$. The reason for this appears to be that the annealing rates in zircon are influenced by the amount of α-damage, namely the damage done to the crystal lattice by recoil of nuclei as they emit α particles in the decay chain of uranium (Garver and Kamp, 2002). Zircons that exhibit significant amounts of α-damage anneal much more rapidly and have much lower effective closure temperatures than do zero-damage zircons (Kasuya and Naeser, 1988; Yamada *et al.*, 1995; Rahn *et al.*, 2004).

The effects of annealing can be quantified by measuring the lengths of horizontal confined tracks (Gleadow *et al.*, 1986), i.e. tracks parallel to the polished face of the grain that do not cut the surface but have been etched because they intersect cracks or other tracks allowing access to the etchant (see Figures 3.6 and 3.7). Because tracks are formed continuously, each track experiences a different portion of the integrated thermal history. Thus, the track-length distribution, which is obtained by measuring the lengths of a sufficient number of confined tracks (preferably ≥ 100), contains information on the thermal history experienced by the sample. The track-length distribution therefore provides valuable additional constraints on the interpretation of fission-track ages for samples that have experienced prolonged or complex thermal histories. Track-length measurements are at present standard practice in apatite fission-track thermochronology whenever the sample character and track density make this feasible.

Annealing models

The kinetics of fission-track annealing in apatite have been evaluated empirically by laboratory experiments (Green *et al.*, 1986; Crowley *et al.*, 1991; Carlson *et al.*, 1999), enabling a quantitative determination of the relationship between thermal history and track length distribution (Laslett *et al.*, 1987; Green *et al.*, 1989b; Crowley *et al.*, 1991; Gallagher, 1995; Ketcham *et al.*, 1999).

The annealing of fission tracks is a temperature-dependent diffusional process to which the Arrhenius law can be applied. However, in contrast to the diffusion models for isotopic methods, there is no accepted physical model of fission-track annealing processes at the atomic level, although Carlson (1990) attempted to derive one. The reason for this is that the process of fission-track annealing is much more complicated than the diffusion of a single atomic species out of a mineral lattice. Fission tracks are made up of multiple defects, each with their own activation energy for elimination (Green et al., 1988; Wagner and Van den Haute, 1992; Ketcham et al., 1999). Moreover, what we are seeing under the optical microscope are *revealed* tracks, so the interaction between the mineral defects and the chemical etching agent also plays a role. Qualitatively, fission tracks anneal at first by tip shortening before they become segmented by the development of unetchable gaps between the segments (Green et al., 1986; Carlson, 1990). For this reason, models of fission-track annealing have been developed using an empirical approach, looking at what form of the annealing relationship best fits the data statistically, rather than an *ab initio* physical approach. The most general form of an empirical annealing model can be written as

$$g(r; a, b) = f(t, T; C_i) \qquad (3.14)$$

where r indicates the degree of annealing (usually quantified as the track-length reduction l/l_0, where l is the measured mean track length and l_0 the initial track length), t is time, T is temperature and a, b and C_i are fitting parameters.

The empirical nature of the annealing models produces a degree of ambiguity in the interpretation of the laboratory data. Thus, several functional forms for the model have been proposed. For the right-hand side, a 'parallel Arrhenius' model can be derived, for which contours of equal annealing r form parallel lines in an Arrhenius plot:

$$f = C_0 + C_1 \ln(t) + \frac{C_2}{T} \qquad (3.15)$$

This equation is a specific form of the general Arrhenius relationship (2.9). A subtle but statistically significant improvement in the fit to the laboratory annealing data is obtained by using a 'fanning Arrhenius' model, in which contours of equal r fan out from a single point (Laslett et al., 1987; Crowley et al., 1991; Ketcham et al., 1999):

$$f = C_0 + C_1 \frac{\ln(t) + C_2}{1/T - C_3} \qquad (3.16)$$

Ketcham et al. (1999) required their model to fit both their laboratory annealing data and two 'benchmarks' for annealing on geological timescales and suggested that the best fit was obtained with a 'curvilinear Arrhenius' model in which contours of equal annealing are slightly curved (Figure 3.9):

$$f = C_0 + C_1 \frac{\ln(t) + C_2}{\ln(1/T) - C_3} \qquad (3.17)$$

As shown in Figure 3.9, although the fits of these different functional forms of the annealing model to the laboratory data are nearly indistinguishable, once they

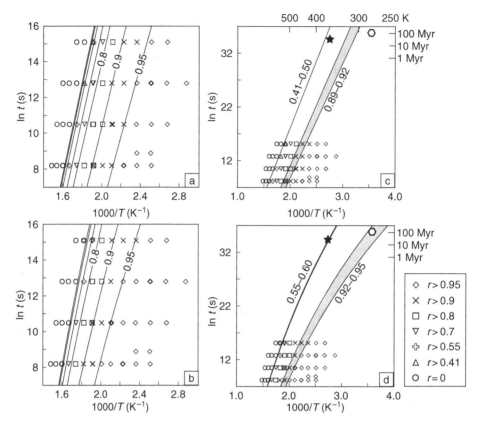

Fig. 3.9. Fission-track annealing data for apatite and their extrapolation to geological time scales. In (a) and (b) are shown the laboratory annealing data of Carlson et al. (1999) for end-member fluor-apatite; the relative track-length reduction r in different experiments is plotted as a function of the logarithm of time and inverse temperature. Lines of equal track-length reduction r are plotted for the fanning Arrhenius model (Equation (3.16)) fit of Ketcham et al. (1999) in (a) and the fanning curvilinear (Equation (3.17)) fit in (b). In (c) and (d) are shown comparisons between model predictions and two geological 'benchmarks': the black star indicates a strongly annealed high-temperature sample (the deepest sample recovered from Flaxman's well in the Otway Basin (Gleadow and Duddy, 1981), cf. Figure 3.8); the open hexagon indicates a practically unannealed low-temperature sample (recovered from a drill-core collected during Ocean Drilling Program leg 129 (Vrolijk et al., 1992)). Shaded lines indicate the $T-t$ conditions predicted for the amount of track-length reduction r recorded in these samples; the difference in r-values between the plots stems from the fact that in (c) the raw track-length data are plotted whereas in (d) these are projected onto the C-axis. Modified from Ketcham et al. (1999). Reproduced with permission by the Mineralogical Society of America.

3.3 Fission-track thermochronology

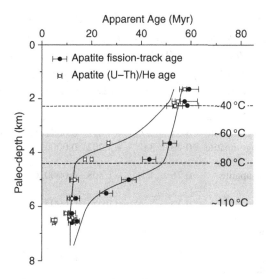

Fig. 3.10. Apatite fission-track and (U–Th)/He ages as a function of initial depth in the crust for samples from the White Mountains, California, a crustal block that was rapidly exhumed ∼12 Myr ago along a major extensional detachment. Data are from Stockli et al. (2000). Dashed lines indicate approximate locations of the top and bottom of the apatite He partial-retention zone at 40 and 80 °C, respectively; the grey-shaded region represents the apatite fission-track partial-annealing zone between ∼60 and 110 °C. The data are consistent with a pre-exhumational history of stability since 55 Myr ago along a geothermal gradient of 15 ± 2 °C km^{-1}; continuous lines show predicted apatite fission-track and (U–Th)/He ages for such a scenario (predictions are made using MadTrax and MadHe codes, the Laslett et al. (1987) algorithm for fission-track annealing and a grain size of 65 μm for He diffusion). Modified from Farley (2002). Reproduced with permission by the Mineralogical Society of America.

are extrapolated over many orders of magnitude to geological timescales, the model predictions differ significantly, the 'fanning Arrhenius' model predicting more rapid annealing on geological timescales than the other models. The models do, however, provide acceptable fits to ages obtained from samples with well-constrained thermal histories (e.g., Corrigan, 1993) (Figure 3.10).

For the left-hand side of Equation (3.14), most authors (Laslett et al., 1987; Crowley et al., 1991; Ketcham et al., 1999) have used a Box–Cox transform:

$$g(r; a, b) = \frac{\left[(1 - r^b)/b\right]^a - 1}{a} \qquad (3.18)$$

A simpler form was proposed by Laslett and Galbraith (1996):

$$g(r) = \ln(1 - r) \qquad (3.19)$$

Laboratory parameterisations of the model have been published by Laslett et al. (1987) for Durango apatite ([Cl]/[Cl + F] = 0.2), Crowley et al. (1991) for

Table 3.5. *Parameters for commonly used annealing models. For the apatite models, parameters refer to Equations (3.18) and (3.16) except for the curvilinear fit of Ketcham et al. (1999), which refers to Equation (3.17). For the zircon models, parameters refer to Equation (3.21).*

Model	Composition	a	b	C_0	C_1	C_2	C_3
Laslett (1987)	Durango apatite	0.35	2.7	−4.87	0.000168	28.095	0
Crowley et al. (1991)	Durango apatite	0.49	3.0	−3.202	0.0000937	2.567	0.0004200
Crowley et al. (1991)	Fluor-apatite	0.76	4.16	−1.508	0.00002076	8.581	0.0009967
Ketcham et al. (1999)	Apatite (fanning)	0.15	8.76	−11.053	0.0003896	17.842	0.0006767
Ketcham et al. (1999)	Apatite (curvilinear)	0.20	7.42	−26.039	0.5317	62.319	−7.8935
Tagami et al. (1998)	α-Damaged zircon			11.41	0.0002472	0.01125	
Rahn et al. (2004)	Zero-damage zircon			11.57	0.0002755	0.01075	

Durango apatite and fluor-apatite ([Cl]/[Cl+F] = 0.02) and Ketcham et al. (1999) for a range of apatite compositions. Values for the parameters in Equations (3.16)–(3.18) fitted to these experimental data are given in Table 3.5.

In the multi-compositional model of Ketcham et al. (1999), the actual track-length shortening for any apatite composition is related to that of the most resistant apatite in their study by

$$r_{lr} = \left(\frac{r_{mr} - r_{mr_0}}{1 - r_{mr_0}} \right)^k \quad (3.20)$$

where r_{lr} and r_{mr} are the reduced lengths of the less-resistant and more-resistant apatites, respectively, and r_{mr_0} is the reduced length of the most resistant apatite at the point where the less-resistant apatite becomes totally annealed. Ketcham et al. (1999) provided a calibration of the fitting parameters r_{mr_0} and k with some of the most common measures of apatite composition, such as Cl^- content and etch-pit width.

Fewer experiments have been conducted on zircon, due to the much longer time required to produce equivalent degrees of annealing in this mineral (see Rahn et al. (2004) for a review); models of the form (3.15) and (3.16) have been fitted to these data (Yamada et al., 1995; Galbraith and Laslett, 1997; Tagami et al., 1998; Rahn et al., 2004). In Table 3.5 we reproduce the parameter fits from the

3.3 Fission-track thermochronology

last two of these studies, for α-damaged and zero-damage zircons, respectively. They have been fitted to a simplified annealing equation:

$$\ln(1 - r) = C_0 + C_1 T \ln t + C_2 T \tag{3.21}$$

MadTrax.f

The subroutine MadTrax.f models the response of fission tracks to an arbitrary input thermal history. This is provided by the user in the form of a series of temperature–time pairs, and the subroutine calculates the annealing experienced by tracks formed throughout the sample's history, integrating the results to produce a synthetic fission-track age and length distribution. The forward model used in this subroutine is described in detail in Appendix 1.

MadTrax.f is based on the annealing models proposed by Laslett *et al.* (1987) and Crowley *et al.* (1991); the user determines what model parameters to use. A user-friendly forward and inverse model of fission-track annealing, based on the 'multi-kinetic' model of Ketcham *et al.* (1999), is provided by the code AFTSolve. An academic, non-commercial version of this code, developed by Ketcham *et al.* (2000), is available by downloading from http://www.apatite.com/AFTSolve.html.

Tutorial 2

Use the codes Muscovite.f90 and MadTrax.f to predict muscovite ^{40}Ar/^{39}Ar ages and apatite fission-track ages and length distributions, respectively, for the five time–Temperature histories enumerated in Tutorial 1 (Chapter 2). Compare these with the (U–Th)/He ages predicted previously. What do you deduce from these comparisons, in terms of the discriminating power of (combinations of) different thermochronometers?

4

The general heat-transport equation

In this chapter, we derive the differential equation governing the transport of heat in solids. This equation is used to estimate the contributions from conduction, tectonic advection and radiogenic heat production in geological systems. We discuss the various types of boundary conditions that are applicable in the context of tectonic and geomorphic problems. We then provide the solution of the heat-transport equation under the assumption of one-dimensionality and neglecting the effect of rock advection towards the cold surface. In doing so we provide a reference conductive solution under a range of assumptions concerning the conductivity and the rate of heat production in the crust.

4.1 Heat transport within the Earth

In solids, heat is principally transported by conduction or advection. Conduction implies a transfer of molecular vibrational energy; advection implies a spatial reorganisation of the internal heat of the system by the relative translation of some of its parts.

Within the Earth, heat is constantly being produced by the natural radioactive decay of unstable isotopes, principally uranium, potassium and thorium. The heat is produced at such a rate and over such a large volume (the entire crust and mantle) that, if it were transported out of the system by conduction only, the entire planet would rapidly reach melting temperature. In fact, the transport is mostly by advection: most of the Earth's mantle is at such a temperature that it is sufficiently weak to flow over geological timescales. Following Archimedes' principle, the hotter, thus less dense, parts of the Earth's interior tend to rise towards the surface; in doing so they transport heat by advection. This buoyancy-driven advection is called convection. As rocks move towards the cold surface, they cool, becoming more rigid, and conduction progressively takes over as the

4.2 Conservation of energy

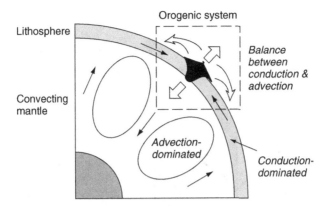

Fig. 4.1. Modes of heat transport in the Earth. In the convecting mantle, advection dominates, whereas within the lithosphere, which can be regarded as a thermal boundary layer, conduction dominates, except in active tectonic areas (orogenic systems), where both conduction and advection operate. Note that the thickness of the lithosphere (≈ 100 km) has greatly been exaggerated compared with the mantle thickness (2900 km).

dominant heat-transport mechanism. The relatively thin region over which this conductive transfer takes place is called the lithosphere.

Being at the interface between the hot, solid Earth and the relatively cold overlying atmosphere/hydrosphere, the lithosphere is the coldest part of the Earth system and thus the strongest. This outer shell is composed of a series of discrete, almost rigid plates that are in relative horizontal motion with respect to each other. These relative motions are driven by the underlying large-scale convective flow and may cause the plates to deform, especially along their edges. For example, where two plates collide, they are subjected to compressional forces, which, locally, may lead to thickening and uplift of the surface (Figure 4.1). The topographic gradient induced by the localised uplift causes erosion, transport and deposition of rocks, which, in turn, may cause transport of heat by advection. Thus, within the conducting lithosphere, the transport of heat may become locally dominated by advection.

4.2 Conservation of energy

Let's consider the heat balance within an infinitesimal volume of material as shown in Figure 4.2. At each point of space, we can define the flux of heat per unit area and unit time, $\vec{q} = [q_x, q_y, q_z]$. The amount of heat entering the system per unit time through one side of the volume is thus given by the product of the component of the heat flux perpendicular to that side and its surface area. If we

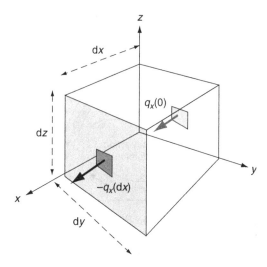

Fig. 4.2. Heat balance in the x-direction. The difference between the heat flux entering the infinitesimal volume of size $dx \times dy \times dz$ and that leaving the volume must be compensated by an increase in temperature.

consider one of the two faces perpendicular to the x-axis, we can thus write that the heat entering the system through that face per unit time is given by

$$Q_x = q_x \, dy \, dz \qquad (4.1)$$

The net amount of heat, ΔQ_x, entering the small volume per unit time in the x-direction is given by the difference between the heat entering and the heat leaving through each of the two faces perpendicular to the x-axis:

$$\begin{aligned} \Delta Q_x &= -Q_x(dx) + Q_x(0) \\ &= -(q_x(dx) - q(0)) dy \, dz \\ &= -\frac{q_x(dx) - q_x(0)}{dx} dx \, dy \, dz \end{aligned} \qquad (4.2)$$

Adding the contributions from each of the other two directions leads to

$$\begin{aligned} \Delta Q &= +\Delta Q_x + \Delta Q_y + \Delta Q_z \\ &= -\left(\frac{q_x(dx) - q_x(0)}{dx} + \frac{q_y(dy) - q_y(0)}{dy} + \frac{q_z(dz) - q_z(0)}{dz} \right) dx \, dy \, dz \end{aligned} \qquad (4.3)$$

The heat capacity, c, of a given system is defined as the amount of heat that must be given to a unit mass of the system to increase its temperature by one unit of temperature over a unit of time. By virtue of conservation of energy, we can write

$$\rho c \frac{dT}{dt} dx\,dy\,dz = \Delta Q \tag{4.4}$$

where ρ is the density of the material, and thus

$$\rho c \frac{dT}{dt} = -\left(\frac{q_x(dx) - q_x(0)}{dx} + \frac{q_y(dy) - q_y(0)}{dy} + \frac{q_z(dz) - q_z(0)}{dz} \right) \tag{4.5}$$

We can express the conservation of energy at each point of the system by collapsing the sides of the small infinitesimal volume to zero, i.e. by calculating the limit of the right-hand side of (4.5) for dx, dy and $dz \to 0$, which leads to

$$\rho c \frac{dT}{dt} = -\frac{\partial q_x}{dx} - \frac{\partial q_y}{dy} - \frac{\partial q_z}{dz} = -\mathrm{div}\cdot \vec{q} \tag{4.6}$$

where $\mathrm{div}\cdot \vec{q}$ is the *divergence* of the heat flux, that is the sum of the partial derivatives of the components of the flux. We can thus state that, at any point of the system, the rate of change of temperature is proportional to the divergence of the heat flux.

4.3 Conduction

The conduction of heat obeys Fourier's law, which states that the conductive heat flux per unit area, $\vec{q}$, is proportional to the local temperature gradient, $\mathrm{grad}\,T = [\partial T/\partial x, \partial T/\partial y, \partial T/\partial z]$. The constant of proportionality is called the conductivity of the material, k. Heat always propagates from warm to cold regions; thus the flux of heat must be in the direction opposite to that of the temperature gradient. This leads to the following form of *Fourier's law for heat conduction*:

$$\vec{q} = -k\,\mathrm{grad}\,T \tag{4.7}$$

Introducing this expression into (4.6) leads to the following partial differential equation governing the transient transport of heat by conduction:

$$\rho c \frac{dT}{dt} = \mathrm{div}\cdot k\,\mathrm{grad}\,T \tag{4.8}$$

In cases in which the conductivity does not vary spatially, Equation (4.8) becomes

$$\rho c \frac{dT}{dt} = k \nabla^2 T \tag{4.9}$$

where

$$\nabla^2 T = \frac{\partial^2 T}{\partial x^2} + \frac{\partial^2 T}{\partial y^2} + \frac{\partial^2 T}{\partial z^2} \qquad (4.10)$$

is called the *Laplacian operator*.

Consequently, the basic equation governing transient conductive heat transport takes the general form of a diffusion equation, i.e. the local rate of change of temperature is proportional to the second spatial derivative or 'curvature' of the temperature field.

4.4 Advection

In the above basic equation (4.10), the spatial coordinates $[x, y, z]$ are attached to the small volume of rock for which the heat balance is derived. This frame of reference is somewhat unconventional in Earth-science problems, with the system of reference more commonly being attached to a particular point in space – the centre of the Earth, or some arbitrary location on the Earth's surface (Figure 4.3). Material points (rocks) may move with respect to this system of reference, for

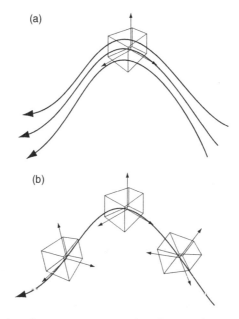

Fig. 4.3. Two basic reference systems used to describe heat transport in deforming systems: (a) the Eulerian system of reference, in which the observer (i.e. the spatial coordinates) is fixed in space and considers the change in temperature in an infinitesimally small volume through which rock is advected at a known rate determined by the velocity field $[v_x, v_y, v_z]$; and (b) the Lagrangian system of reference, in which the observer is attached to the moving/deforming system.

example by convective motion in the mantle or by erosion and exhumation in the crust. If we wish to use spatial rather than material coordinates, we need to modify the basic conductive-heat-transport equation.

At a given location $[x, y, z]$, the rate of change of temperature is not just equal to the local rate of change of temperature in the material (the rocks); we must also consider the fact that rocks are advected at a velocity $\vec{v}$ and that there might exist a temperature gradient in the material in that direction. In this case the temperature at location $[x, y, z]$ will change at a rate that is equal to the product of the velocity and the temperature gradient, such that the total rate of change of temperature is given by

$$\frac{dT}{dt} = \frac{\partial T}{\partial t} + v_x\frac{\partial T}{\partial x} + v_y\frac{\partial T}{\partial y} + v_z\frac{\partial T}{\partial z} = \frac{\partial T}{\partial t} + \vec{v}\cdot\text{grad } T \quad (4.11)$$

One may also state that, in an Eulerian description of heat transport, i.e. where spatial coordinates are fixed, space and time cannot be considered as independent variables; consequently, the rate of change of temperature is equal to the total derivative of temperature with respect to time and we must write

$$\frac{dT}{dt} = \frac{\partial T}{\partial t} + \frac{\partial T}{\partial x}\frac{\partial x}{\partial t} + \frac{\partial T}{\partial y}\frac{\partial y}{\partial t} + \frac{\partial T}{\partial z}\frac{\partial z}{\partial t}$$

$$= \frac{\partial T}{\partial t} + v_x\frac{\partial T}{\partial x} + v_y\frac{\partial T}{\partial y} + v_z\frac{\partial T}{\partial z} = \frac{\partial T}{\partial t} + \vec{v}\cdot\text{grad } T \quad (4.12)$$

On combining this result with Equation (4.10), we obtain the general transient heat-transport equation for conduction and advection:

$$\rho c\left(\frac{\partial T}{\partial t} + \vec{v}\cdot\text{grad } T\right) = k\nabla^2 T \quad (4.13)$$

4.5 Production

In the Earth, most rocks contain a finite concentration of radioactive isotopes. In terms of their overall contribution of heat to the crust and mantle, the most important of these are isotopes of U, Th and K. The decay of these radioactive atoms into their daughter products is accompanied by an infinitesimal loss in mass, which must be compensated by an increased kinetic energy of the newly created particles and, potentially, by a radiative energy flux. This additional energy must be taken into account in the global heat budget.

If we define H as the rate of radiogenic heat production per unit mass, the rate of heat produced by unit volume is ρH. Adding this term to the heat-balance equation leads to

$$\rho c\left(\frac{\partial T}{\partial t} + \vec{v}\cdot\text{grad } T\right) = k\nabla^2 T + \rho H \quad (4.14)$$

4.6 The general heat-transport equation

In its explicit form, the general heat-transport/balance equation in the solid Earth can thus be written as

$$\rho c \left(\frac{\partial T}{\partial t} + v_x \frac{\partial T}{\partial x} + v_y \frac{\partial T}{\partial y} + v_z \frac{\partial T}{\partial z} \right) = \frac{\partial}{\partial x} k \frac{\partial T}{\partial x} + \frac{\partial}{\partial y} k \frac{\partial T}{\partial y} + \frac{\partial}{\partial z} k \frac{\partial T}{\partial z} + \rho H \quad (4.15)$$

where T is the temperature, t is time, x, y and z are the three spatial coordinates and v_x, v_y and v_z are the corresponding components of rock velocity (in a very general way, defined with respect to the centre of the Earth, or, in a more practical way, with respect to the surface of the Earth), k is the thermal conductivity, ρ is the density, c is the heat capacity and H is the rate of radioactive heat production per unit volume. Further details on the derivation of this equation may be found in Carslaw and Jaeger (1959) or Turcotte and Schubert (1982).

4.7 Boundary conditions

To understand the thermal structure of the Earth's crust and its evolution through time, one must find a solution to this equation or, most probably, one of its simplified forms that also conforms to a set of boundary conditions and, in the transient case, to an initial temperature distribution. Boundary conditions are essentially of two types; they correspond to either a fixed temperature (Dirichlet-type boundary condition),

$$T = T_S \quad (4.16)$$

which is usually imposed at the surface or at the base of the crust, or a fixed conductive heat flux, in a direction normal to the boundary (Neumann-type boundary condition),

$$k \frac{\partial T}{\partial z} = q_m \quad (4.17)$$

which is usually applied at the base of the crust and/or along the vertical side boundaries of the region of interest. Note that, for practical reasons, we will now assume that the z-axis is positive downwards (as shown in Figure 4.4), and thus vertical heat flux will usually be positive. This is because the temperature usually increases with depth; temperature gradients are thus positive and so are conductive heat fluxes. Other types of boundary conditions could be considered (radiative and convective for instance) but are not covered here.

As an example, we show in Figure 4.4 the boundary conditions commonly used in crustal-tectonics problems: fixed temperature along the free surface, fixed heat flux along the base, representing the heat loss by conduction from the

4.7 Boundary conditions

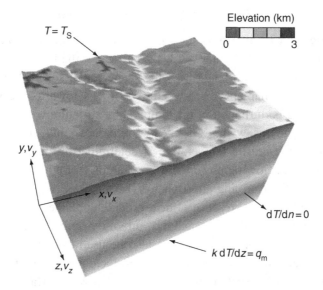

Fig. 4.4. The general problem of determining the temperature structure within the crust including the effect of finite relief amplitude at the surface and tectonic advection at a rate v_x, v_y and v_z. Usually one assumes a fixed temperature at the surface of the Earth and a constant heat flux from the underlying mantle. The side boundaries are isolated conductively.

underlying mantle and, if one considers the problem in more than one dimension, no conductive heat loss through the side boundaries. The last condition imposes that the temperature gradient in a direction normal to the 'side walls' must be zero, unless heat is transported by advection through those walls, which is commonly avoided by assuming that the advection velocity vector has no component in a direction normal to the walls.

Note too that a fixed temperature condition does not need to be uniform in space; for example, when temperature calculations are limited to the Earth's crust, where the range of temperature is of the order of a few hundred degrees, one may need to take into account that the mean surface temperature is closer to 15 than to 0 °C. If one's purpose is to determine the temperature history of rocks in order to calculate their cooling age for a low-temperature thermochronometer such as (U–Th)/He in apatite, one will need to take into account the atmospheric lapse rate, i.e. the rate of change of temperature with altitude.

It is also worth mentioning that a 'fixed-temperature' boundary condition does not need to be constant in time. For example, one may wish to study the rate and depth of penetration of the temperature anomaly created by the diurnal or yearly cycle in atmospheric temperature by imposing a cyclic variation in temperature at the Earth's surface (see Turcotte and Schubert (1982) for example).

4.8 Purely conductive heat transport

As stated earlier, our main purpose in using thermochronology is to constrain the rate at which rocks are exhumed towards the surface of the Earth and, potentially, to estimate the rate at which landforms (the shape of the surface) evolve. As we will see in the next chapters, the advection of rocks towards the surface may lead to a noticeable perturbation of the temperature structure within the crust; but before we investigate this, we must define the 'reference', undisturbed, conductive solution whereby the crustal temperature is determined by the one-dimensional balance between heat flowing upwards by conduction from the underlying mantle and heat lost into the overlying atmosphere. This is a rather simple problem that is complicated only by the presence of heat-producing elements such as uranium, thorium and potassium in the crust and the variability of the thermal conductivity of crustal rocks. This simple solution can also be used to determine the thermal history of rocks being exhumed in regions of low tectonic activity, typically where exhumation is driven by isostatic rebound associated with surface erosion.

Conductive equilibrium – uniform conductivity – no heat production

The simplest form of the heat-transfer equation assumes steady state ($\partial T/\partial t = 0$), a purely conductive heat transport in the vertical dimension, a fixed surface temperature, $T(z=0) = T_S$, and, at the base of the crust, a fixed heat flux from the mantle

$$k \frac{\partial T}{\partial z}(z=L) = q_m$$

In this situation, the general heat-transfer equation (4.15) becomes

$$\frac{\partial^2 T}{\partial z^2} = 0 \qquad (4.18)$$

which, after integration, can be further reduced to

$$k \frac{\partial T}{\partial z} = q_m \qquad (4.19)$$

Integrating (4.19) leads to a simple linear increase of temperature with depth:

$$T(z) = T_S + \frac{q_m}{k} z \qquad (4.20)$$

Thus, in a tectonically quiet region where rocks have a low content of radiogenic heat-producing elements and where conductivity (which we will assume is constant within, and can therefore be read as a proxy for, a given lithology) is uniform, the temperature increases linearly with depth. Typical mantle heat-flow values are in the range 10–30 mW m^{-2}; a commonly accepted value for the mean crustal conductivity is 3Wm^{-1}K^{-1} (Turcotte and Schubert, 1982). The rate of increase of temperature with depth should therefore be in the range 3–10 °C km^{-1}. In most regions of the Earth, however, the geothermal gradient is significantly higher, typically of the order of 15–20 °C km^{-1} (Turcotte and Schubert, 1982). This is due to the contribution to the heat balance from heat-producing elements (see Section 4.8).

In a region where exhumation is very slow (typically <0.1 mm yr^{-1}), conduction dominates heat transport. Assuming that the contribution from heat-producing elements can be neglected, the conductive temperature distribution given by (4.20) is a good approximation that can be used to determine the exhumation rate, $\dot{E}$, from thermochronological data. If t_c is the age of a rock for a thermochronometric system characterised by a closure temperature T_c, one can write

$$T_c = T_S + \frac{q_m}{k}\dot{E}t_c \qquad (4.21)$$

or, assuming a zero surface temperature,

$$\dot{E} = \frac{T_c}{Gt_c} \qquad (4.22)$$

where G is the conductive geothermal gradient, $G = q_m/k$.

Equation (4.22) shows that, to determine the exhumation rate under the assumption of conductive equilibrium, one needs to know the mantle heat flow which, in this case, is equivalent to the local surface geothermal gradient divided by the conductivity of the rock. This is true even in the case when two or more ages (i.e. time–temperature pairs) are known.

Conductive equilibrium – variable conductivity

As shown in Table 4.1, the thermal conductivity of rocks is highly variable. From Equation (4.15), it is clear that spatial variations in thermal conductivity will lead to spatial variability in temperature. We will now show that, even in situations where the Earth's crust could be assumed to be laterally homogeneous and to have reached conductive equilibrium, vertical variations in thermal conductivity lead to changes in vertical temperature gradient.

For a material in which thermal conductivity varies with depth, the equation of conductive thermal equilibrium can be written as

$$\frac{\partial}{\partial z}k(z)\frac{\partial T}{\partial z} = 0 \qquad (4.23)$$

Under the assumption that the crust is made of a series of horizontal layers, each characterised by a different conductivity, at equilibrium, the temperature must increase linearly within each layer. The geotherm is thus made of a series of linear

Table 4.1. *Thermal conductivity for a range of rock types; data from Clauser and Huenges (1995)*

Rock type	Conductivity (W m^{-1} K^{-1})
Sedimentary rocks	0.5–4.5 (inversely $\propto$ porosity)
Volcanic rocks	1.5–3.5 (inversely $\propto$ porosity)
Metamorphic rocks (low quartz content)	1–4
Metamorphic rocks (high quartz content)	5–7

segments of varying slope, inversely proportional to the thermal conductivity, such that the heat flow through each layer, $q_0 = q_m$, is uniform with depth and equal to the product of the conductivity of the layer, k_i, and the local geothermal gradient:

$$q_0 = -k_i \frac{\partial T}{\partial z} \qquad (4.24)$$

Regions where rocks have a low thermal conductivity (such as in sedimentary basins) are characterised by a high geothermal gradient, whereas regions where rocks have a high thermal conductivity (such as in high-grade metamorphic terranes) are characterised by a relatively low geothermal gradient.

The exact form of the temperature distribution can be written as

$$T(z) = A_i \frac{z - d_i}{l_i} + B_i \qquad \text{for } d_i < z < d_{i+1} = d_i + l_i \qquad (4.25)$$

where l_i is the thickness of layer i and d_i the depth to the top of layer i. The value of the $2N$ constants A_i and B_i can be derived from the assumed temperature boundary condition at the surface,

$$T(z = 0) = T_S \qquad (4.26)$$

the imposed mantle heat flux at the base,

$$k_N \frac{\partial T}{\partial z}\bigg|_{z=d_N+l_N} = q_m \qquad (4.27)$$

and the continuity of temperature and conservation of energy (continuity of heat flux) along interfaces between each pair of layers.

The surface and bottom boundary conditions give us

$$B_1 = T_S$$

$$A_N = \frac{q_m l_N}{k_N} \qquad (4.28)$$

while continuity of temperature leads to

$$A_i + B_i = B_{i+1} \quad \text{for } i = 1, N-1 \quad (4.29)$$

and continuity of heat flux to

$$\frac{k_i A_i}{l_i} = \frac{k_{i+1} A_{i+1}}{l_{i+1}} \quad \text{for } i = 1, N-1 \quad (4.30)$$

Combining these relationships leads to the following expressions for the A_i and B_i:

$$A_i = \frac{l_i q_m}{k_i}$$
$$B_i = q_m \sum_{j=1}^{i-1} \frac{l_j}{k_j} \quad (4.31)$$

If we now assume that rocks are being exhumed at a low but constant rate through such a thermal structure, they will experience several periods of uniform cooling, corresponding to the exhumation of each layer, as shown in Figure 4.5. Naïve application of Equation (4.22), in the absence of constraints on regional thermal-conductivity variations, may lead to the interpretation of rapid (and wholly artefactual) changes in exhumation rate having occurred. It is therefore important to take into account any knowledge one may have as to the conductivity structure of the crust during exhumation in the interpretation of thermochronological data.

In general, most sediment types are characterised by a relatively low conductivity such that sedimentary basins are usually characterised by a high geothermal gradient; this is called the 'blanketing effect' of the sediments on the underlying crust (see Sandiford (1999) for an interesting discussion of the effect of the sedimentary cover on the thermal structure of the crust and its rheological implications).

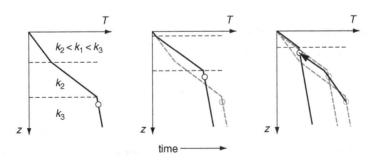

Fig. 4.5. The thermal history of a rock particle exhumed within a vertically layered conductivity structure.

Table 4.2. *Heat production for a range of rock types; data from Förster and Förster (2000)*

Rock type	Heat production ($\mu W\ m^{-3}$)
Sedimentary rocks	0.5–5.5
Granitic rocks	2.5–3.5
	(varies with differentiation)
Metamorphic rocks	1.5–3.5

In a study of the denudation history of the southeastern Brazilian margin, Gallagher *et al.* (1994) showed how failure to take the nature of the overburden removed by denudation (in their case, low-conductivity volcanics) into account may lead to erroneous estimates of the depth of denudation from apatite fission-track (FT) data.

Conductive equilibrium – the effect of heat production

Plate-tectonic behaviour can be viewed as a response to heat transport in a thermally stratified system. Heat produced by radioactive isotopes keeps the interior of the Earth hot enough to drive convective overturn of the mantle, producing movement in the overlying coupled lithospheric plates. Many of these elements exist in the mantle but at relatively low concentrations. During the formation and differentiation of the Earth and the subsequent formation of the continental crust by magmatic processes, these elements became concentrated in the continental crust. The presence of radioactive elements and the heat they produce perturb the conductive thermal gradient.

Typical values of the heat produced by various rocks are given in Table 4.2. Heat-producing elements are most concentrated in granitic rocks and the sediments derived from them.

The one-dimensional steady-state conductive-heat-transport equation including the effect of heat production is

$$k\frac{\partial^2 T}{\partial z^2} + \rho H = 0 \qquad (4.32)$$

Assuming a uniform distribution of radioactive elements in the crust, the general solution of this equation is a second-order polynomial (or quadratic function):

$$T = a_0 + a_1 z + a_2 z^2 \qquad (4.33)$$

4.8 Purely conductive heat transport

By introducing this solution into the differential equation, one can easily show that

$$a_2 = -\frac{\rho H}{2k} \quad (4.34)$$

Assuming that the surface temperature is fixed, $T(z=0) = T_S$, and that a constant (mantle) heat flux, q_m, is imposed at the base of the crust, i.e. at $z = L$, one can easily find the values of the two other constants, a_0 and a_1:

$$a_0 = T_S$$
$$a_1 = \frac{q_m}{k} + \frac{\rho H L}{k} \quad (4.35)$$

such that the temperature distribution within the crust is given by

$$T(z) = -\frac{\rho H}{2k} z^2 + \left(\frac{q_m}{k} + \frac{\rho H L}{k}\right) z + T_S \quad (4.36)$$

This expression can be used to derive the temperature history, $T(t)$, of a rock particle being exhumed at a constant and slow rate, $\dot{E}$:

$$T(t) = -\frac{\rho H}{2k} (\dot{E}t)^2 + \left(\frac{q_m}{k} + \frac{\rho H L}{k}\right) \dot{E}t + T_S \quad (4.37)$$

One can readily see that the thermal history of a rock particle being exhumed in a region characterised by a finite and uniform concentration of heat-producing elements is quadratic in time (i.e. it varies as the square of time), predicting an accelerated cooling in the recent past and thus an apparent increase in exhumation rate if temperature is wrongly interpreted as a simple, linear proxy for depth (see Figure 4.6).

If we have at least two independent age–temperature constraints (corresponding to two different thermochronological systems), we can derive the values of the two coefficients in Equation (4.37),

$$\frac{\rho H \dot{E}^2}{2k} \quad \text{and} \quad \left(\frac{q_m}{k} + \frac{\rho H L}{k}\right) \dot{E} \quad (4.38)$$

assuming that we know the surface temperature, T_S. We can then extract the value of the exhumation rate, $\dot{E}$, by assuming that we know either the mantle component of the heat flow, q_m, or the mean heat-production rate, ρH.

Note that because heat-producing elements are mostly concentrated in the upper crust, one commonly assumes that heat production is limited to a layer in the uppermost part of the crust of thickness l. In this case, the temperature distribution in the crust is made of two separate domains: the top one (i.e. from $z = 0$ to $z = l$) is characterised by a quadratic increase of temperature with depth given by

$$T(z) = -\frac{\rho H}{2k} z^2 + \left(\frac{q_m}{k} + \frac{\rho H l}{k}\right) z + T_S \quad (4.39)$$

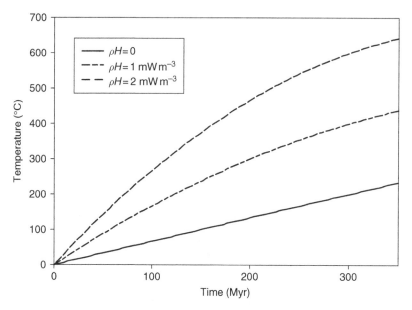

Fig. 4.6. Predicted thermal histories for a rock particle experiencing exhumation at a rate of 100 m Myr^{-1} for a period of 350 Myr. The conductivity is 3 W m^{-1} K^{-1}, the crustal thickness is 35 km and the mantle heat flux is 20 mW m^{-2}. The three curves correspond to different values for the assumed heat production.

and the bottom one is characterised by a linear increase of temperature with depth at a rate that is given by the ratio of the mantle heat flux and the conductivity:

$$T(z) = T_S + \frac{\rho H l^2}{2k} + \frac{q_m}{k} z \qquad (4.40)$$

Note that, at $z = l$, the two solutions lead to the same expression for the temperature ($T = T_S + \rho H l^2/(2k) + q_m l/k$) and heat flux ($q = k\, \partial T/\partial z = q_m$).

A similar expression can be derived in situations where the layer of assumed uniform heat production of thickness l is buried at a depth h beneath the surface (see Appendix 3).

Another common assumption is that the concentration of heat-producing elements in the crust decays exponentially with depth. In this case the one-dimensional, steady-state heat equation becomes:

$$k \frac{\partial^2 T}{\partial z^2} + \rho H_0 e^{-z/z_0} = 0 \qquad (4.41)$$

where H_0 is the rate of heat production at the surface and z_0 is the depth over which this rate of production decreases by a factor e. Assuming a constant surface

4.8 Purely conductive heat transport

Table 4.3. *Surface heat-flow and heat-production measurements from South Australia (from Neumann et al. (2000))*

Heat flow (mW m^{-2})	Heat production (μW m^{-3})
48	2.7
71	3.1
76	3.8
57	4.9
109	7.4
127	7.9
92	8.2

temperature and a constant mantle heat flux at the base of the crust (i.e. at $z = L$), one can show that the solution to that equation is (Turcotte and Schubert, 1982)

$$T(z) = T_S + \left(\frac{q_m}{k} - \frac{\rho H_0 z_0}{k} e^{-L/z_0}\right) z - \frac{\rho H_0 z_0^2}{k}\left(e^{-z/z_0} - 1\right) \quad (4.42)$$

Tutorial 3

(a) Use the dataset shown in Table 4.3 to derive the mantle component to surface heat flow in this area of South Australia. (b) Assuming that the heat-producing elements are uniformly distributed in a thin layer located in the uppermost crust, what is the thickness of this layer? (c) Predict the one-dimensional thermal structure in this region of the Earth's crust and (d) derive the thermal history of a rock being exhumed at a rate of $\dot{E} = 0.1\,\text{km}\,\text{Myr}^{-1}$ assuming that exhumation has no effect on the distribution of heat-producing elements. Hint: use the relationship $q_0 = q_m + \rho H L$, relating the surface heat flux, q_0, to the mantle heat flux, q_m, the heat-production rate, ρH, and the thickness of the heat-producing layer, L.

5
Thermal effects of exhumation

> *In our quest to extract information on exhumation from thermochronological data, we now turn our attention to the effects that exhumation itself, i.e. the advection of rocks towards the Earth's surface, may have on the temperature structure of the crust. It is important to realise that the process that we are trying to quantify, namely rock exhumation, is strongly non-linear: it perturbs the system that we propose to use to measure it, i.e. the temperature history of the rock.*

5.1 Steady-state solution

We first provide a means of determining the perturbation caused by advection under the assumption of steady-state exhumation, i.e. for situations in which rock uplift is balanced by erosion over a long period of time. Although this is an idealised scenario, it is worth considering, in order to assess the magnitude of the perturbation and the conditions under which it is likely to be significant. Furthermore, thermal steady state is not an unlikely scenario when considered on the scale of an entire, mature orogen, such as in the Southern Alps of New Zealand or in the Taiwan orogen (Willett and Brandon, 2002).

Uplift and exhumation

As already mentioned in Section 1.2, it is very important always to keep in mind that the cooling that is documented by thermochronology during exhumation is related to the advection of rocks towards the cold upper surface of the Earth. Our ability to derive geological constraints from thermochronology is limited to a spatial window between a sample's location at the time of sampling and a depth range corresponding to a temperature range centred around the closure temperature of

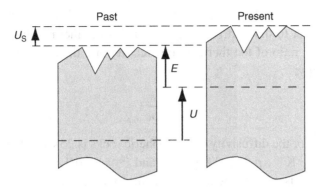

Fig. 5.1. U is rock uplift, i.e. the vertical movement of rocks with respect to the centre of the Earth; U_S is surface uplift, i.e. the vertical movement of the surface of the Earth with respect to the centre of the Earth; E is exhumation, i.e. the movement of rocks relative to the surface of the Earth. $E = U - U_S$ (see also Figure 1.6 for a more detailed description of these important concepts).

the relevant system, i.e. comprised between the first retention (or "open-system") and blocking temperatures. As illustrated in Figure 5.1, thermochronology provides information on rock exhumation only; it does not provide constraints on either rock or surface uplift (England and Molnar, 1990; Brown, 1991).

Note also that, because the surface of the Earth is relatively planar, the horizontal components of the temperature gradient are much smaller than the vertical one. In most situations, horizontal tectonic movements do not result in measurable changes in temperature and, therefore, thermochronology cannot be used directly to constrain horizontal rock movement. We will discuss the consequences of exceptions to this rule in Chapter 10.

In this chapter, we will therefore focus on the one-dimensional, vertical-advection case; in Chapters 6 and 7 we will consider the two- and three-dimensional cases, respectively.

Basic PDE: the steady-state case

The simplest way to interpret thermochronological data is to assume that an equilibrium between tectonic rock uplift and erosion has been reached, which leads to a constant exhumation rate, $\dot{E}$. If this situation is maintained for a long period of time, a thermal steady state develops, which is governed by a balance between advection and conduction (Willett and Brandon, 2002). In this situation, the temperature, T, within the crust does not change with time and is governed by the following simplified form of Equation (4.15):

$$-\dot{E}\frac{\partial T}{\partial z} = \kappa \frac{\partial^2 T}{\partial z^2} \tag{5.1}$$

where $\dot{E}$ is the exhumation rate, i.e. the vertical component of the velocity of rocks with respect to the Earth's surface, and κ is the thermal diffusivity, which is defined as the ratio of the thermal conductivity and the product of the density and heat capacity:

$$\kappa = \frac{k}{\rho c} \quad (5.2)$$

A typical value for the diffusivity of lithospheric rocks is $25\,\text{km}^2\,\text{Myr}^{-1}$ (assuming that $k \approx 2\,\text{W}\,\text{m}^{-1}\,\text{K}^{-1}$, $\rho \approx 3000\,\text{kg}\,\text{m}^{-3}$ and $c \approx 1000\,\text{W}\,\text{K}^{-1}\,\text{kg}^{-1}$). As previously assumed, z is zero at the surface of the Earth and is positive downwards. Thus, the exhumation rate is negative because it represents a negative velocity (i.e. in the direction opposite to z). We assume that the surface temperature is constant:

$$T(z=0) = 0 \quad (5.3)$$

and we consider the situation in which temperature is fixed at some depth L in the Earth:

$$T(z=L) = T_L \quad (5.4)$$

The dimensionless form

It is useful to derive the dimensionless form of this equation by introducing a new depth coordinate normalised by the position of the lower boundary, $z' = z/L$, and a new temperature, normalised by the temperature at the lower boundary, $T' = T/T_L$. The equation becomes

$$-\frac{\dot{E}L}{\kappa}\frac{\partial T'}{\partial z'} = \frac{\partial^2 T'}{\partial z'^2} \quad (5.5)$$

This equation can be further simplified by introducing the Péclet number, Pe, defined as $Pe = \dot{E}L/\kappa$. The equation and boundary conditions become

$$-Pe\frac{\partial T'}{\partial z'} = \frac{\partial^2 T'}{\partial z'^2}$$
$$T'(0) = 0 \quad (5.6)$$
$$T'(1) = 1$$

The Péclet number is a dimensionless ratio that measures the efficiency of advective versus conductive heat transport. In situations for which $Pe \gg 1$, the system is dominated by advection; where $Pe \ll 1$ the system is dominated by conduction.

5.1 Steady-state solution

The Péclet number can also be regarded as the ratio of two timescales, the timescale for conduction, $\tau_c = L^2/\kappa$, and the timescale for advection, $\tau_a = L/\dot{E}$:

$$Pe = \tau_c/\tau_a = \frac{\dot{E}L}{\kappa} \tag{5.7}$$

In situations for which conductive transport is more efficient than advective transport, the timescale for conduction is smaller than the timescale for advection, and Pe is small; when advection dominates, the timescale for advection is smaller and Pe is large.

Solution

The solution of the dimensionless form of equation (5.6) is

$$T'(z') = \frac{1 - e^{-Pe z'}}{1 - e^{-Pe}} \tag{5.8}$$

It is shown in dimensionless-variable space for a range of values of the Péclet number in Figure 5.2(a).

The most important point to notice from the form of the solution (Equation (5.8)) is how heat advection by exhumation affects the distribution of temperature within the crust: the linear conductive temperature distribution (Equation (4.21)) is transformed into an exponential increase of temperature with depth. This clearly

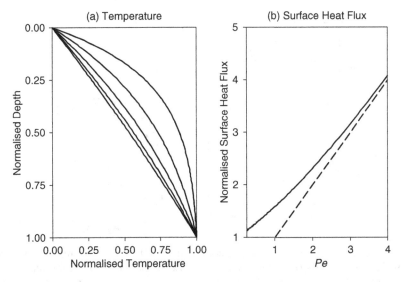

Fig. 5.2. (a) Geotherms calculated for a range of values of the Péclet number. Pe values are 0.25, 0.5, 1, 2 and 4. (b) The normalised surface heat flux as a function of Pe (the dashed line indicates the 1:1 relationship).

illustrates one of the main difficulties that arises when using thermochronology to determine exhumation rate: ages correspond to the time when a rock, usually now at the surface of the Earth, cooled through a given temperature; to determine the exhumation rate we must determine the depth of this isotherm which, in turn, is a function of the exhumation rate. In short, the age of a rock, t_c, is a non-linear function of the exhumation rate. In the simplified situation envisaged here, this non-linear relationship can be inverted to give

$$t_c = -\frac{\kappa}{\dot{E}^2} \ln\left(1 - \frac{T_c}{T_0}(1 - e^{-\dot{E}L/\kappa})\right) \quad (5.9)$$

where T_c is the closure temperature.

The second important point to stress is how a simplistic interpretation of thermochronological data can lead to erroneous conclusions about the exhumation history of a rock particle. For a finite value of the Péclet number, Equation (5.8) shows that cooling accelerates exponentially as the rock approaches the surface. This situation is illustrated in Figure 5.3. If one neglects the perturbation induced by advection on the vertical temperature structure of the crust, one is mistakenly led to conclude that the exhumation rate has progressively increased in the recent past.

Finally, it is worth noting that a reasonably accurate value of the Péclet number can be derived from a simple geophysical measurement. Indeed, using

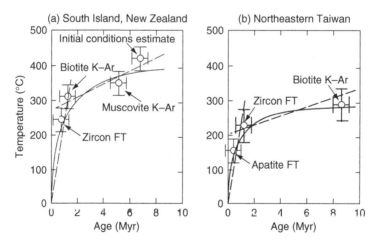

Fig. 5.3. Multi-system thermochronological data from (a) Kamp and Tippett (1993) and (b) Lan et al. (1990), showing how an increase in cooling rate can be interpreted as an indication of a recent change in exhumation rate (dashed lines). In both cases, the data is equally well explained by a constant exhumation rate leading to an exponential decrease in temperature with time (solid lines) (Batt and Braun, 1997).

Equation (5.6), one can derive the following expression for the dimensionless surface temperature gradient:

$$\left.\frac{\partial T'}{\partial z'}\right|_{z'=0} = \frac{Pe}{1-e^{-Pe}} \qquad (5.10)$$

which tends towards 1 as the Péclet number tends towards zero. Thus the surface-heat-flux anomaly between a tectonically active region (characterised by a finite value of Pe) and the neighbouring tectonically quiet region (where $Pe=0$) is given by

$$\frac{\left.k\frac{\partial T}{\partial z}\right|_{Pe}}{\left.k\frac{\partial T}{\partial z}\right|_{Pe=0}} = \frac{Pe}{1-e^{-Pe}}$$

which, for large values of the Péclet number, i.e. $Pe \geq 2$, tends towards Pe (see Figure 5.2(b)). Thus, by measuring surface heat flow in an orogenic area and normalising it with respect to a similar measurement made in the vicinity of the orogen, one obtains a direct estimate of the Péclet number characterising the orogen and, subsequently, an estimate of the curvature of the geothermal gradient in response to erosional advection.

5.2 Thermal effects of exhumation: transient solution

Steady-state conditions are encountered in many orogenic settings, in part due to the steadiness of relative plate motion at the Earth's surface. However, discrete events punctuate the life of orogenic systems, typically corresponding to plate reorganisation events or major climatic changes.

We now ask the following question: how can we quantify the effect of rapid changes in exhumation on the thermal structure of the crust in order to extract information on the timing of such events from thermochronological datasets? Although many complex scenarios can be envisaged, we will focus here on situations characterised by a sudden increase in exhumation rate that may be related to the initiation of a major tectonic event or that is associated with rapid climate change.

Basic PDE: the transient case

To study the effect of a rapid change in exhumation rate, one needs to consider the transient solution to the heat-transfer equation, which is given by

$$\rho c \left(\frac{\partial T}{\partial t} - \dot{E}\frac{\partial T}{\partial z} \right) = \frac{\partial}{\partial z} k \frac{\partial T}{\partial z} + \rho H \qquad (5.11)$$

where ρ is density, c is heat capacity, $\dot{E}$ is exhumation rate, k is conductivity and H is heat production per unit mass. We assume that the temperature at the surface is always zero, $T(z=0, t) = 0$ and neglect heat production ($H = 0$). We further assume that the system is coming out of a period of slow tectonic activity ($\dot{E} \approx 0$), such that the initial temperature profile is linear, $T(z, t=0) = Gz$, where G is the conductive geothermal gradient. We consider now how the temperature structure is going to evolve when, at $t = 0$, the exhumation rate becomes finite.

Solution

The solution to this equation can be found in Carslaw and Jaeger (1959):

$$T(z, t) = G(z + \dot{E}t) + \frac{G}{2}\left[(z - \dot{E}t)e^{-\dot{E}z/\kappa}\operatorname{erfc}\left(\frac{z - \dot{E}t}{2\sqrt{\kappa t}}\right)\right.$$

$$\left. - (z + \dot{E}t)\operatorname{erfc}\left(\frac{z + \dot{E}t}{2\sqrt{\kappa t}}\right)\right] \quad (5.12)$$

where κ is the thermal diffusivity. The solution involves terms containing the complementary error function which is defined by the following integral expression:

$$\operatorname{erfc} x = \frac{2}{\sqrt{\pi}} \int_x^\infty e^{-u^2} du \quad (5.13)$$

Although this function now appears as an intrinsic function in most Fortran and C compilers, we provide a method by which to evaluate it in Appendix 7.

For convenience's sake, we define θ as $\theta = t_0 - t$, where t_0 is the time at which the exhumation/tectonic episode started. For rocks that are now (i.e. at $\theta = 0$) at the surface of the Earth, we can state that

$$z(\theta) = \dot{E}\theta \quad (5.14)$$

On substituting this into Equation (5.12), one can obtain an expression for the temperature history of surface rocks:

$$T(\theta) = \frac{G\dot{E}}{2}\left[(2\theta - t_0)e^{-\dot{E}^2\theta/\kappa}\operatorname{erfc}\left(\frac{\dot{E}(2\theta - t_0)}{2\sqrt{\kappa(t_0 - \theta)}}\right)\right.$$

$$\left. - t_0\operatorname{erfc}\left(\frac{\dot{E}t_0}{2\sqrt{\kappa(t_0 - \theta)}}\right) + 2t_0\right] \quad (5.15)$$

The dimensionless form of the solution

If we scale temperature by the initial temperature of the rock, $T' = T/T_0$, where $T_0 = G\dot{E}t_0$, and time by the time at which exhumation started, $t' = \theta/t_0$, one can express the temperature history in its dimensionless form:

$$T' = \frac{1}{2}\left[(2t'-1)e^{-Pe^2 t'}\operatorname{erfc}\left(\frac{Pe(2t'-1)}{2\sqrt{1-t'}}\right) - \operatorname{erfc}\left(\frac{Pe}{2\sqrt{1-t'}}\right) + 2\right] \quad (5.16)$$

It is interesting to note that, for a rock now at the surface of the Earth, its temperature history normalised by its initial temperature (Equation (5.16)), depends again on a single dimensionless quantity, Pe, the Péclet number, when expressed in terms of time in the past normalised by the time of initiation of exhumation.

Assuming that the mean thermal diffusivity of crustal rocks is known, the temperature history described by Equation (5.15) or (5.16) depends on five parameters: T_0, $\dot{E}$, t_0, g and z_0, of which only three are independent, since we know that $T_0 = Gz_0$ and $z_0 = \dot{E}t_0$. From joint consideration of a finite number of age–closure-temperature pairs, one can extract constraints on these three independent parameters (see Moore and England, (2001)).

5.3 Thermal effects of exhumation: the general transient problem

Tectonic processes are usually complex in their geometrical and temporal evolution. Although analytical solutions are very useful to understand the first-order behaviour of a system, they cannot be used in many real problems. For example, one may wish to estimate the temperature evolution of a system whose exhumation rate is known to vary with time. In such situations, one usually calls upon numerical methods to solve the heat-transport equation.

Here we will briefly describe how the general vertical-heat-transport equation

$$\rho c \frac{\partial T}{\partial t} - \dot{E}\frac{\partial T}{\partial z} = \frac{\partial}{\partial z}k\frac{\partial T}{\partial z} + \rho H$$

$$T = T_S \quad \text{on } S_1 \quad (5.17)$$

$$k\frac{\partial T}{\partial z} = q_S \quad \text{on } S_2$$

can be solved by using the finite-element method. This is a very flexible numerical method that is particularly well suited for the solution of problems involving complex geometries, spatially and temporally varying parameters and multi-dimensional analyses.

In this section, the reader will be given the opportunity to understand this approach in some detail, as we proceed through each of the steps from the transformation of the partial differential equation into an integral equation and its discretisation in terms of finite elements, to the solution of the large system of algebraic equations that this leads to. We cannot offer to cover completely the field of the finite-element method, or even its application to solving the diffusion–advection equation. Many books and textbooks have been written on the subject (Zienkiewicz, 1977; Bathe, 1982). What we offer here is designed to inform the reader about the method and provide enough information to allow him or her either to understand the one-dimensional finite-element code that we provide with this textbook or to write a code that is better suited to his/her needs.

To facilitate the comprehension of rather complex mathematical developments, we will use the one-dimensional (vertical) version of the heat-transport equation to illustrate the finite-element method. It is clear, however, that it is in multi-dimensional problems that the finite-element method is the most powerful. The generalisation from one to two or three dimensions is relatively straightforward and we will provide a few hints on how it is done.

Finite-element equations

The weak form

Assuming that we wish to solve this differential equation in a region of space Ω, one can state that the differential equation is equivalent to (i.e. has the same solution as) the following integral equation:

$$\int_\Omega \psi \left[\rho c \left(\frac{\partial T}{\partial t} + \dot{E} \frac{\partial T}{\partial z} \right) - \frac{\partial}{\partial z} k \frac{\partial T}{\partial z} - \rho H \right] dz + \oint_{S_2} \psi' \left(k \frac{\partial T}{\partial n} - q_S \right) dS = 0$$
(5.18)

where ψ and ψ' are arbitrary functions that have a 'reasonable' form (for example, they do not become infinite inside the domain of integration). We assume that T has been selected such that all Dirichlet-type conditions (on S_1) are automatically satisfied (see Equation (5.17)). Non-homogeneous conditions will be treated later, by modifying the global algebraic system of equations resulting from the finite-element equations directly. S_2 is the part of the boundary of the domain Ω on which Neumann-type boundary conditions (or flux boundary conditions) are imposed.

5.3 Thermal effects of exhumation: the general transient problem

Through integration by parts, one can write

$$\int_\Omega \left[\rho c \left(\psi \frac{\partial T}{\partial t} + \psi \dot{E} \frac{\partial T}{\partial z} \right) + \frac{\partial \psi}{\partial z} k \frac{\partial T}{\partial z} - \psi \rho H \right] dz$$
$$- \oint_S \psi k \left(\frac{\partial T}{\partial z} n_z \right) dS + \oint_{S_2} \psi' \left(k \frac{\partial T}{\partial n} - q_S \right) dS = 0 \quad (5.19)$$

and, without loss of generality, one can assume that

$$\psi' = \psi \quad (5.20)$$

such that the integral equation simplifies to

$$\int_\Omega \left[\rho c \left(\psi \frac{\partial T}{\partial t} + \psi \dot{E} \frac{\partial T}{\partial z} \right) + \frac{\partial \psi}{\partial z} k \frac{\partial T}{\partial z} - \psi \rho H \right] dz - \oint_{S_2} \psi q_S dS = 0 \quad (5.21)$$

under the assumption that a 'no-flux' boundary condition applies on the parts of S that are not included in S_1 or S_2. The homogeneous Neumann-type boundary conditions are also called 'natural' boundary conditions for the finite-element method because they do not require additional treatment.

The weighted residual

The finite-element approximation requires that the volume of integration be discretised, or broken down into a number of discrete elements of smaller volume, across which the differential equation can be accurately solved. To do this, we approximate the unknown, T, by the following expansion:

$$T(x, z) \approx \tilde{T}(x, z) = \sum_1^n N_i(x, y) T_i = \mathbf{NT} \quad (5.22)$$

where the T_i are the approximate values of the unknown field T at a finite number, n, of points or nodes. The N_i are called *shape functions*. They are used to define the numerical or finite-element solution (an approximate representation of the true solution) everywhere from a finite number of nodal values.

In the integral form of the equation, we can now introduce, in place of any function, ψ, a finite, arbitrary set of prescribed functions, ψ_i, for $i = 1, \ldots, n$, where n is the number of nodes. This leads to a set of n integral equations for n unknowns:

$$\sum_i \left\{ \int_\Omega \left[\rho c \left(\psi_j N_i \frac{\partial T_i}{\partial t} + \psi_j \dot{E} \frac{\partial N_i}{\partial z} T_i \right) + \frac{\partial \psi_j}{\partial z} k \frac{\partial N_i}{\partial z} T_i - \psi_j \rho H \right] dz \right.$$
$$\left. - \oint_{S_2} \psi_j q_S dS \right\} = 0 \quad (5.23)$$

The Galerkin method

Many solutions are possible for the choice of the functions ψ_j, including $\psi_j = \delta_j$, called the point-collocation method, or $\psi_j = 1$ in Ω_j, called the sub-domain collocation. The most widely used method is the Galerkin method, in which $\psi_j = N_j$.

Under this assumption, the discretised integral equation takes the form of a set of n integral equations for n unknowns:

$$\sum_i \left\{ \int_\Omega \left[\rho c \left(N_j N_i \frac{\partial T_i}{\partial t} + N_j \dot{E} \frac{\partial N_i}{\partial z} T_i \right) + \frac{\partial N_j}{\partial z} k \frac{\partial N_i}{\partial z} T_i - N_j \rho H \right] dz \right.$$

$$\left. - \oint_{S_2} N_j q_S dS \right\} = 0 \tag{5.24}$$

Finite-element discretisation

The volume Ω is divided into n_e elements, Ω_e. The division is space filling (Figure 5.4) such that the union of the elemental volumes is the total volume of integration,

$$\cup \Omega_e = \Omega \tag{5.25}$$

and the intersection of the elements is empty,

$$\cap \Omega_e = \emptyset \tag{5.26}$$

This division is then used to build the shape function, N_i, one for each element, $N_i^e(x, y)$.

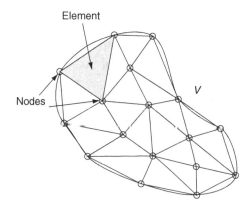

Fig. 5.4. An example of nodes and elements used to discretise an integration volume, V.

5.3 Thermal effects of exhumation: the general transient problem

The finite-element equations can be written in the following matrix form:

$$M \frac{\partial \mathbf{T}}{\partial t} + V\mathbf{T} + K\mathbf{T} = Q \qquad (5.27)$$

where M is the mass matrix, V is the advection matrix, K is the conduction matrix and Q is the production matrix. The components of these matrices are given by the following expressions:

$$\begin{aligned} M_{ij} &= \sum_e M^e_{ij} = \sum_e \int_{\Omega_e} \rho c N_i N_j \, dz \\ K_{ij} &= \sum_e K^e_{ij} = \sum_e \int_{\Omega_e} k \left(\frac{\partial N_i}{\partial z} \frac{\partial N_j}{\partial z} \right) dz \\ V_{ij} &= \sum_e V^e_{ij} = \sum_e \int_{\Omega_e} \rho c \left(\dot{E} N_i \frac{\partial N_j}{\partial z} \right) dz \\ Q_i &= \sum_e Q^e_i = \sum_e \left(\int_{\Omega_e} N_i \rho H \, dx \, dz + \oint_{S_e} N_i q_S \, dS \right) \end{aligned} \qquad (5.28)$$

$\mathbf{T}$ is the solution vector of the nodal temperature, T_i.

Shape functions

A wide variety of shape functions can be used. These can be complex, but are usually chosen to be simple polynomial expressions (see Braun and Sambridge (1995), for example). These shape functions are commonly constructed element by element, i.e. the coefficients of the polynomial expressions are different within each element and depend on the geometry of the element (the coordinates of the nodes defining the elements). The number of nodes per element can also vary.

The general expression defining an elemental shape function is

$$T(x, z) = \sum_{i=1}^{m} N_i(x, z) T_i = N_i T_i \qquad (5.29)$$

where m is the number of nodes in the element.

Note that shape functions are orthogonal,

$$N_i(x_j, z_j) = \delta_{ij} \qquad (5.30)$$

and most finite-element implementations are based on the assumption that the shape functions are iso-parametric:

$$z = N_i z_i \qquad (5.31)$$

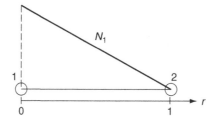

Fig. 5.5. A two-node element and the linear shape function, N_1.

Linear shape functions

The simplest shape function is a polynomial function of order 1:

$$N_i(z) = a_i + b_i z \tag{5.32}$$

Such a shape function is defined on a two-node line (see Figure 5.5).

One can express the coefficients in terms of the nodal coordinates, $[z_i]$:

$$a_1 = \frac{z_2}{z_2 - z_1} \qquad a_2 = \frac{z_1}{z_1 - z_2}$$
$$b_1 = \frac{1}{z_1 - z_2} \qquad b_2 = \frac{1}{z_2 - z_1} \tag{5.33}$$

Local coordinates

It is often more convenient to define shape functions by using a local coordinate system $[r]$. For the case of the linear two-node element and for the particular set of local coordinates defined in Figure 5.5, this leads to

$$N_1 = 1 - r$$
$$N_2 = r \tag{5.34}$$

If one uses local coordinates, one needs to perform the proper change of variables in the various integral expressions:

$$\int_{z_1}^{z_2} f(z) dz = \int_0^1 f(r) \det \boldsymbol{J} \, dr \tag{5.35}$$

where $\boldsymbol{J}$ is the *Jacobian* of the coordinate transformation between $[z]$ and $[r]$:

$$\boldsymbol{J} = \left[\frac{\partial z}{\partial r} \right] \tag{5.36}$$

Note that, in one dimension, the Jacobian is a scalar; in two dimensions, it is a 2×2 matrix; and in n dimensions, it is an $n \times n$ matrix. The main advantages of the local coordinates are that they provide (a) a more straightforward definition of the shape functions and (b) a more practical definition of the integration points used in the numerical estimation of the various integrals, as we shall show later.

5.3 Thermal effects of exhumation: the general transient problem

Quadratic shape functions

To solve a second-order differential equation (as is the case for the heat-transport equation), one usually uses quadratic shape functions defined on a three-node element (Figure 5.6). In local coordinates, the shape functions are defined by

$$N_1 = (2r-1)(r-1)$$
$$N_2 = r(2r-1) \quad (5.37)$$
$$N_3 = 4r(1-r)$$

Two-dimensional elements

Commonly used two-dimensional elements are the linear (three-node), bi-linear (four-node), quadratic (six-node) and bi-quadratic (eight-node) elements. Their respective shape functions are

$$N_1 = 1 - r - s$$
$$N_2 = r \quad (5.38)$$
$$N_3 = s$$

in the case of the three-node triangular element,

$$N_1 = \frac{1}{4}(1+r)(1+s)$$
$$N_2 = \frac{1}{4}(1-r)(1+s)$$
$$N_3 = \frac{1}{4}(1-r)(1-s) \quad (5.39)$$
$$N_4 = \frac{1}{4}(1+r)(1-s)$$

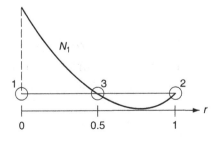

Fig. 5.6. A three-node element and the quadratic shape function, N_1.

in the case of the four-node rectangular element,

$$N_1 = (1 - r - s) - \frac{1}{2}N_4 - \frac{1}{2}N_6$$

$$N_2 = r - \frac{1}{2}N_4 - \frac{1}{2}N_5$$

$$N_3 = s - \frac{1}{2}N_5 - \frac{1}{2}N_6 \qquad (5.40)$$

$$N_4 = 4r(1 - r - s)$$

$$N_5 = 4rs$$

$$N_6 = s(1 - r - s)$$

in the case of the six-node triangular element and

$$N_1 = \frac{1}{4}(1+r)(1+s) - \frac{1}{2}N_5 - \frac{1}{2}N_8$$

$$N_2 = \frac{1}{4}(1-r)(1+s) - \frac{1}{2}N_5 - \frac{1}{2}N_6$$

$$N_3 = \frac{1}{4}(1-r)(1-s) - \frac{1}{2}N_6 - \frac{1}{2}N_7$$

$$N_4 = \frac{1}{4}(1+r)(1-s) - \frac{1}{2}N_7 - \frac{1}{2}N_8 \qquad (5.41)$$

$$N_5 = \frac{1}{2}(1-r^2)(1+s)$$

$$N_6 = \frac{1}{2}(1-s^2)(1-r)$$

$$N_7 = \frac{1}{2}(1-r^2)(1-s)$$

$$N_8 = \frac{1}{2}(1-s^2)(1+r)$$

in the case of the eight-node element. Their respective geometries are shown in Figure 5.7.

Shape-function derivatives

The shape-function derivatives are easily estimated when global coordinates are used; for example, in the case of linear, two-node elements

$$\frac{\partial N_i}{\partial z} = b_i \qquad (5.42)$$

where the b_i are defined in (5.33).

5.3 Thermal effects of exhumation: the general transient problem

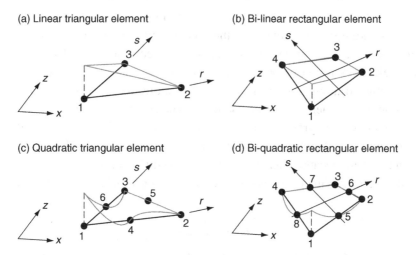

Fig. 5.7. Two-dimensional finite elements and the shape function N_1.

Using local coordinates, the shape-function spatial derivatives are more complex to define and require the computation of the inverse of the transformation matrix. For example, in two dimensions, the derivatives of the shape functions are estimated from

$$\begin{bmatrix} \dfrac{\partial}{\partial x} \\ \dfrac{\partial}{\partial z} \end{bmatrix} = \begin{bmatrix} \dfrac{\partial r}{\partial x} & \dfrac{\partial s}{\partial x} \\ \dfrac{\partial r}{\partial z} & \dfrac{\partial s}{\partial z} \end{bmatrix} \begin{bmatrix} \dfrac{\partial}{\partial r} \\ \dfrac{\partial}{\partial s} \end{bmatrix} = \dfrac{1}{J} \begin{bmatrix} \dfrac{\partial x}{\partial r} & \dfrac{\partial z}{\partial r} \\ \dfrac{\partial x}{\partial s} & \dfrac{\partial z}{\partial s} \end{bmatrix} \begin{bmatrix} \dfrac{\partial}{\partial r} \\ \dfrac{\partial}{\partial s} \end{bmatrix} \quad (5.43)$$

where J, the Jacobian, is the determinant of

$$\begin{vmatrix} \dfrac{\partial x}{\partial r} & \dfrac{\partial y}{\partial r} \\ \dfrac{\partial x}{\partial s} & \dfrac{\partial z}{\partial s} \end{vmatrix}$$

Which elements to use?

The selection of an element type is governed by several factors, including the nature of the partial differential equation to be solved, the nature of the solution, the geometry of the problem, the architecture of the computer on which the equation will be solved, the ease of implementation, etc.

For cases in which the heat-transport equation is dominated by the conduction term (and the temperature is likely to vary as a quadratic function of space during the transient stage and to be a linear function at steady state), quadratic elements (i.e. three-node elements) are most suitable since they will provide an appropriate support for the 'smooth' solution. For cases in which the advection term is dominant (or plays an important role), the solution will 'look like' an exponential function of space and a higher-order element might be necessary; however, if spatial discontinuities are present in the conductivity or any other parameter, the solution (or at least its derivative) is likely to be discontinuous, and a large number of linear elements might be better suited to represent the solution. In two dimensions, whether triangles or rectangles are the more appropriate elements to use depends on the geometry of the problem, i.e. whether it is characterised by a radial geometry, a plane of symmetry, or a curved boundary, for example. It is also possible to combine different types of elements in the same problem but this involves additional coding because each element might be different from its neighbour, and might have a different set of shape functions and/or a different integration rule.

Numerical integration

The construction of the finite-element matrices (Equation (5.28)) requires the computation of spatial integrals. Because the integrands are not simple functions, this integration cannot be performed analytically (i.e., exactly) and, instead, numerical integration is used. This operation consists of estimating the integrand at a finite number of so-called 'integration points' and approximating the integral by a weighted sum of these estimates. Several schemes to determine the positions of the integration points and the values of the weights exist, the most common of which are the *Newton–Cotes* and *Gauss formulas*.

The Newton–Cotes formula

An approximation of the integral

$$\int_a^b F(r) \mathrm{d}r \tag{5.44}$$

can be obtained by estimating the integrand at $n+1$ equally spaced points, r_i, defining n intervals between a and b. The approximate value of the integral is obtained from a weighted sum of these estimates:

$$\int_a^b F(r) \mathrm{d}r = (b-a) \sum_{i=0}^{n} C_i^n F(r_i) + R_n \tag{5.45}$$

The coefficients or weights (C_i^n) are given in Table 5.1.

Gauss quadrature

The most widely used numerical integration is the *Gauss quadrature*, in which the integral is approximated by a weighted sum of estimates of the integrand at unequally spaced locations:

$$\int_a^b F(r) \mathrm{d}r = \alpha_1 F(r_1) + \alpha_2 F(r_2) + \cdots + \alpha_n F(r_n) + R_n \tag{5.46}$$

5.3 Thermal effects of exhumation: the general transient problem

Table 5.1. *Newton–Cotes weights for one-dimensional numerical integration*

n	C_0^n	C_1^n	C_2^n	C_3^n	C_4^n
1	1/2	1/2			
2	1/6	4/6	1/6		
3	1/8	3/8	3/8	1/8	
4	7/90	32/90	12/90	32/90	7/90

Table 5.2. *Sampling points and weights in Gauss–Legendre numerical integration (interval −1 to +1)*

n	r_i	α_i
1	0	2
2	±0.577 350 269 189 626	1
3	±0.774 596 669 241 483	0.555 555 555 555 556
	0	0.888 888 888 888 889
4	±0.861 136 311 594 053	0.347 854 845 137 454
	±0.339 981 043 584 856	0.652 145 154 862 546

where both the weights α_i and the locations of the sampling points, r_i, are variables given in Table 5.2. Note that the calculation of the locations of Gauss points can become cumbersome for large values of n and approximate values are obtained by using Legendre polynomials to determine them.

These formulas are easily generalised to two dimensions. Various numerical integration schemes adapted to triangular elements are given in Table 5.3. Those adapted to quadrilateral elements are given in Table 5.4.

Which integration scheme?

If Gauss quadrature is used, a polynomial of order $2n-1$ is integrated exactly with an integration scheme of order n. In general, one attempts to estimate the order of the function to be integrated (the elements of the various matrices) and use this information to select the integration order. Note that too high an integration order will lead to unnecessary calculations, whereas too low an order may be inaccurate and, sometimes, impossible since it will create so-called 'zero-energy modes' (i.e. oscillating solutions) and a singular matrix (algebraic system) to solve (Bathe, 1982).

However, in practice, it has been found that the use of a reduced numerical integration order (less than what is required to obtain an exact integration) leads, in

Table 5.3. *Gauss integrations over triangular domains*

Order	Location	r	s	w
Three-point		$r_1 = 0.1667$ $r_2 = 0.6667$ $r_3 = r_1$	$s_1 = r_1$ $s_2 = r_1$ $s_3 = r_2$	$w_1 = 0.3333$ $w_2 = w_1$ $w_3 = w_1$
Seven-point		$r_1 = 0.1013$ $r_2 = 0.7974$ $r_3 = r_1$ $r_4 = 0.4701$ $r_5 = r_4$ $r_6 = 0.0597$ $r_7 = 0.3333$	$s_1 = r_1$ $s_2 = r_1$ $s_3 = r_2$ $s_4 = r_6$ $s_5 = r_4$ $s_6 = r_4$ $s_7 = r_7$	$w_1 = 0.1259$ $w_2 = w_1$ $w_3 = w_1$ $w_4 = 0.1324$ $w_5 = w_4$ $w_6 = w_4$ $w_7 = 0.225$

many cases, to improved results (Bathe, 1982). For instance, reduced integration can produce better-conditioned matrices, which leads to faster convergence of an iterative solver. In other cases, it might be beneficial to use selective integration, i.e. to integrate various terms of the equation with different orders of integration.

In short, the best method is to test various integration methods on a problem for which the solution is known or well characterized and find a compromise among accuracy, ease of computation and stability.

Time-stepping algorithms

The general heat-transport equation is an evolution equation, i.e. it involves the time derivative of the temperature. In other words, knowing the temperature at a time t, one tries to determine the temperature at a time $t + \Delta t$, where Δt is called the time step. As seen in Section 2.5 where we derived the equations of a finite-difference scheme to solve the solid-state diffusion equation, the temporal derivative of the temperature can be estimated either from the temperature at time step t, in which case the time-marching procedure is called *explicit*, or from the temperature at time $t + \Delta t$, in which case the procedure is called *implicit*. Implicit procedures are usually always stable and more accurate than explicit ones. However, because it involves the temperature at time $t + \Delta t$, an implicit

5.3 Thermal effects of exhumation: the general transient problem

Table 5.4. *Gauss integrations over quadrilateral domains*

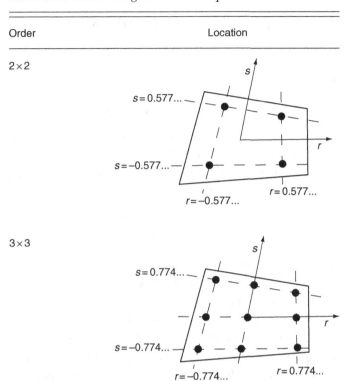

scheme requires a much more complex algebraic operation (the inversion of a matrix) than would an equivalent explicit scheme.

An explicit–implicit scheme

As already mentioned in Section 2.5, the two schemes can be seen as end-members of a more general method in which time integration is performed by combining the values of the time-derivatives of the temperature at times t and $t+\Delta t$. If one assumes that, during the time interval Δt, the temporal derivative of the temperature lies between its values at t and $t+\Delta t$, one can write

$$T(t+\Delta t) \approx T(t) + \Delta t \frac{\partial T}{\partial t}$$

$$= T(t) + \Delta t(1-\alpha)\frac{\partial T}{\partial t}(t) + \Delta t \, \alpha \frac{\partial T}{\partial t}(t+\Delta t) \quad (5.47)$$

If $\alpha = 0$, the scheme is explicit; if $0 < \alpha \leq 1$, the scheme is implicit. One can show (Belytschko *et al.*, 1979) that the most accurate and stable solution is usually obtained with $\alpha = \frac{1}{2}$.

Using this expression, one can re-write the finite-element equations (5.27) as

$$[M + \alpha \Delta t \, K'] \mathbf{T}(t + \Delta t) = [M - (1-\alpha) \Delta t \, K'] \mathbf{T}(t)$$
$$+ \Delta t \, [(1-\alpha) Q(t) + \alpha Q(t + \Delta t)] \quad (5.48)$$

where $K' = K + V$.

Limits on time-step length

The accuracy and stability of the time-integration scheme depends on the length of the time step. As seen in Section 5.1, the heat-transport equation can be characterised by (at least) two timescales: a timescale for conduction, τ_c, and a timescale for advection, τ_a, given by

$$\tau_c = \frac{L^2}{k\rho c} \qquad \tau_a = \frac{L}{\dot{E}} \quad (5.49)$$

where L is the length scale of the problem.

In an explicit scheme, stability and accuracy require that the time step, Δt, be much smaller than the conductive and advective timescales. In an implicit scheme, accuracy requires that the time step be smaller than both timescales. These rather constraining conditions on the length of the time step can be overcome, for example, by using non-Galerkin methods (Hughes and Brooks, 1982).

Stability can be achieved only if the time necessary to transport heat by advection across one element is smaller than the time required to transport heat by conduction. In other words, the elemental Péclet number, defined as

$$Pe_e = \frac{\Delta z \, \dot{E} \rho c}{k} \quad (5.50)$$

must be smaller than unity.

Assembling the matrices

Assembling the finite-element matrices can be difficult, especially if there is more than one degree of freedom (unknown) at each node. In the case of the heat-transport problem, the temperature is the only unknown and the procedure of assembling the matrices is rather simple.

The complete matrix

Let's assume that we have constructed n_e elemental matrices A^e and that we wish to form the global matrix, A. The rank of A^e is $m \times m$, the number of nodes per element, whereas the rank of A is $n \times n$, the total number of nodes. The connectivity between the elements is commonly defined by an array, *icon*,

5.3 Thermal effects of exhumation: the general transient problem

in which $icon(i, e)$ defines the general node number ranging between 1 and n corresponding to node i in element e. Using this notation, the assembly of a global matrix can be expressed as

$$\text{for } e = 1, \ldots, n_e$$
$$\text{for } i = 1, \ldots, m$$
$$\text{for } j = 1, \ldots, m$$
$$\text{do } A_{icon(i,e),icon(j,e)} = A_{icon(i,e),icon(j,e)} + A^e_{i,j} \quad (5.51)$$

Banded matrices and node numbering

The global matrix A is very large and sparse, and consequently requires much memory to store in two and three dimensions. Because of the nature of the finite-element discretisation, global matrices can be made to be diagonally dominant or, even better, banded (Figure 5.8), by appropriately choosing a node-numbering scheme. The distance between element A_{ij} and the diagonal depends on $|i - j|$. The bandwidth of the global matrix can be optimised by selecting a node-ordering scheme that minimises the difference in node numbers between any two nodes belonging to any given element. On a regular or quasi-regular grid, this is relatively easily achieved by a row-by-row or column-by-column numbering scheme. For general grids, automated methods are commonly used (Sloan, 1989).

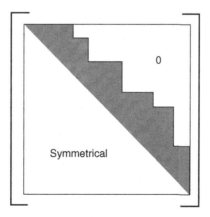

Fig. 5.8. The banded nature of the finite-element matrix. Grey areas correspond to finite (non-nil) components of the matrix; white areas correspond to nil components. With the appropriate node numbering, most finite-element matrices are diagonally dominant and thus characterised by a narrow bandwidth (distance between the off-diagonal components of the matrix and its diagonal).

Skyline storage

The memory required to store the global matrix can be substantially reduced by using a skyline storage scheme. In such a scheme, the matrix is stored as a vector in which the columns of the matrix are stored consecutively (Figure 5.9) and the positions of the diagonal elements of the matrix are stored in a separate index array.

Note that skyline storage is possible only if one uses particular ways of solving the system of algebraic equations, such as the Cholesky factorisation (see below).

Solution of the finite-element equations

The transient finite-element equations (5.48) form a system of algebraic equations in the unknown nodal temperatures, T_i, which can be expressed in a matrix form as

$$A\mathbf{T} = \mathbf{B} \qquad (5.52)$$

Note that the matrix A is symmetrical and positive-definite when advection terms are neglected. This is an important consideration since the nature of the matrix (symmetrical versus asymmetrical) will determine which solution procedure can be used.

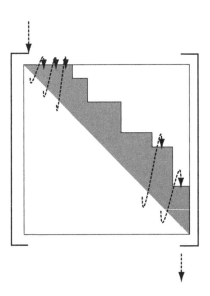

Fig. 5.9. The skyline storage scheme for symmetric matrices. The columns of the matrix are consecutively stored in a linear array, neglecting the nil components of the matrix outside the skyline and thus saving storage.

Fixed-temperature boundary conditions

Several methods by which to impose Dirichlet-type (i.e. fixed-temperature) boundary conditions exist. The most common is based on an a-posteriori modification of the global algebraic system (5.52). If we call T_b the vector containing all fixed (known) nodal temperatures and T_a the vector containing all other, unknown, nodal temperatures, we can re-write the global matrix equation as

$$\begin{bmatrix} A_{aa} & A_{ab} \\ A_{ba} & A_{bb} \end{bmatrix} \begin{bmatrix} T_a \\ T_b \end{bmatrix} = \begin{bmatrix} B_a \\ B_b \end{bmatrix} \tag{5.53}$$

On solving for T_a, we obtain

$$A_{aa} T_a = B_a - A_{ab} T_b \tag{5.54}$$

Consequently, fixed-temperature boundary conditions can easily be imposed by modifying the right-hand-side vector B and global matrix A according to

$$\begin{aligned} B_a &= B_a - A_{ab} T_b \\ B_b &= T_b \\ A_{ab} &= A_{ba} = 0 \\ A_{bb} &= I \end{aligned} \tag{5.55}$$

Positive-definite, symmetrical A

If the matrix A is positive-definite and symmetrical, the solution to the matrix problem can be found by a 'direct method' based on the well-known Cholesky factorisation of A:

$$A = L^T L \tag{5.56}$$

where L is a lower-triangular matrix. Note that this factorisation is possible only when the matrix A is symmetrical and positive-definite because it requires the computation of the square root of the eigenvalues of A. The advection term in the heat-transport equation (5.48) is non-symmetrical; therefore, Cholesky factorisation can be used only when advection is neglected.

The computation of L is performed sequentially according to

$$\left. \begin{aligned} L_{jj} &= \left(A_{ij} - \sum_{k=1}^{j-1} L_{jk}^2 \right)^{1/2} \\ L_{ij} &= \frac{1}{L_{jj}} \left(A_{ij} - \sum_{k=1}^{j-1} L_{ik} L_{kj} \right), \quad \text{for } i = j+1, \ldots, n \end{aligned} \right\} \text{for } j = 1, \ldots, n \tag{5.57}$$

The solution **T** is calculated through a double backsubstitution:

$$L^T C = B$$
$$LT = C \tag{5.58}$$

each of which is a rather simple operation due to the triangular nature of **L**.

Owing to the high performance of modern computers, such a direct approach is now commonly used for most two-dimensional problems. For very large problems (hundreds of thousands of elements) or those in which advection is important, iterative methods are used because they are, in general, more efficient than direct methods, require much less memory and are not limited to solving positive-definite, symmetrical algebraic systems.

Non-symmetrical A

For cases in which **A** is non-symmetrical, the Cholesky factorisation has to be replaced by a more general factorisation of the form

$$A = LU \tag{5.59}$$

where **U** is an upper-triangular matrix. Crout's method is the most efficient means by which to perform this factorisation:

$$\left.\begin{array}{l} L_{11} = 1 \quad U_{11} = A_{11} \\[6pt] U_{ij} = A_{ij} - \sum_{k=1}^{i-1} L_{ik} U_{kj}, \quad \text{for } i = 1, \ldots, j \\[6pt] L_{ij} = \frac{1}{U_{jj}} \left(A_{ij} - \sum_{k=1}^{j-1} L_{ik} U_{kj} \right), \quad \text{for } i = j+1, \ldots, n \end{array}\right\} \text{for } j = 2, \ldots, n \tag{5.60}$$

However, Crout's factorisation is not commonly used to solve large systems of algebraic equations that result from a finite-element solution of a partial differential equation, since it requires pivoting to avoid instabilities associated with computer round-off error. Pivoting means a permutation of the equations to bring the largest possible numbers onto the diagonal; it leads, however, to a reordering of the equations, which, in turn, destroys the banded nature of the matrix.

Iterative methods

For cases in which **A** is non-symmetrical or too large to be solved by a direct method, iterative methods are used. In general, these methods do not require the assembly of the finite-element matrix, **A**, and are therefore much more economical in memory requirements.

5.3 Thermal effects of exhumation: the general transient problem

Most iterative methods are based on the successive improvement of an initial approximation. At each iteration, the solution is updated to reduce the so-called 'residue':

$$\mathbf{R} = \mathbf{B} - A\mathbf{T}^k \qquad (5.61)$$

where $\mathbf{T}^k$ is the approximation to the solution $\mathbf{T}$ at iteration k. The residue can be seen as a measure of how far the approximate solution at iteration k is from the real solution. The key to a good iterative method is to minimise (a) the computational cost of each iteration and (b) the number of iterations. Iterating is usually stopped when the norm of the residue has fallen below a pre-determined value.

Many iterative methods are based on the following scheme:

$$\mathbf{T}^{k+1} = \mathbf{T}^k + \tilde{A}^{-1}(\mathbf{B} - A\mathbf{T}^k) \qquad (5.62)$$

where $\tilde{A}$ is a matrix that 'looks like' A but is much more easily inverted (or factorised). For example, in a Jacobi iterative scheme

$$\tilde{A} = \text{diag } A \qquad (5.63)$$

A commonly used iterative method is the Gauss–Seidel method:

$$\mathbf{T}^{k+1} = \mathbf{T}^k + \beta A_D^{-1}(\mathbf{B} - A_L \mathbf{T}^{k+1} - A_D \mathbf{T}^k - A_U \mathbf{T}^k) \qquad (5.64)$$

where A_L is a lower-triangular matrix, A_D is a diagonal matrix and A_U is an upper-triangular matrix, such that

$$A = A_L + A_D + A_U \qquad (5.65)$$

Note that the rate of convergence of an iterative method is determined by the quality of the approximation that $\tilde{A}$ is of A, but also depends on the conditioning of A. Therefore iterative methods may be unsuitable for some problems in which the matrix is ill-conditioned, such as those dominated by advection (very high Pe).

Most iterative methods require only matrix–vector multiplications and, at most, the computation of the inverse of a diagonal matrix. Therefore, it is not necessary to assemble the complete, large finite-element matrix from the elemental components. Instead, all matrix–vector multiplications can be performed element by element:

$$A\mathbf{T} = \sum_e A_e \mathbf{T} \qquad (5.66)$$

Table 5.5. *Heat1D subroutines and their functions*

Routine	Function
Heat1D.f	Main program
init.f	Where the initial temperature distribution is defined
bc.f	Where the values of the boundary conditions are set
conductivity.f	Where the conductivity is set as a function of time and space
heat.f	Where heat production is set as a function of time and space
velo.f	Where the tectonic exhumation rate is defined as a function of time and space

Heat1D

Included with the notes is a Fortran code named *Heat1D.f*. It is a finite-element code designed to solve the transient heat-conduction equation in one dimension. It uses quadratic elements and allows for spatial and temporal variations in conductivity, heat production and exhumation velocity. Various types of boundary conditions are allowed on either side of the computational domain; and any initial temperature distribution can be considered.

Heat1D is made of a main program, *Heat1D.f*, a set of routines, *init.f*, *bc.f*, *conductivity.f*, *velo.f* and *heat.f* and a set of utility routines (from LAPACK), namely *sgbfa.f*, *sgbsl.f*, *isamax.f*, *saxpy.f*, *sdot.f* and *sscal.f*. All need to be compiled and linked to produce an executable *Heat1D.exe*.

A brief description of each subroutine is given in Table 5.5. All routines can be edited and changed to define the problem to be solved. General description of the problem should be set in *Heat1D.f*, including the geometry, type of boundary conditions and time-step length. Thermal and tectonic properties (initial temperature conditions, conductivity, heat production and velocity) are set in *init.f*, *conductivity.f*, *heat.f* and *velo.f*, respectively.

Comparison with analytical solutions

When developing a numerical algorithm, it is always essential to check the accuracy of the solution it provides against analytical solutions. In Figure 5.10, we compare numerical and analytical solutions to the one-dimensional, transient, conductive heat-transport equation wherein a linear, conductive gradient is perturbed (at time $t = 0$) by a sudden doubling in basal temperature (from $T(z = 1) = 0.5$ to $T(z = 1) = 1$). The solutions are shown at two time steps, $t = 0.05$ and $t = 0.25$, after the thermal perturbation has been introduced and are practically identical,

5.3 Thermal effects of exhumation: the general transient problem

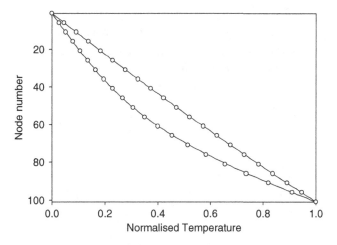

Fig. 5.10. A comparison between analytical (continuous line) and finite-element (open circles) solutions for a transient, heat-conduction problem – see the text for a description of the problem.

i.e. identical within round-off error. The analytical solution is given (in its dimensionless form) in Appendix 5.

Tutorial 4

The thermochronological dataset shown in Table 5.6 was collected at a single location in an active mountain belt, within the hangingwall of and a few hundred metres away from the main thrust which, from seismic-reflection data, is known to extend to a depth of 15 km. What is the approximate maximum temperature reached by this rock during the most recent tectonic event which, from independent data, is known to have started in the Eocene (i.e., 54–38 Myr ago)? Can we derive a mean exhumation rate from this thermochronological dataset? (Assume a value of $25 \, \text{km}^2 \, \text{Myr}^{-1}$ for the thermal diffusivity of crustal rocks).

Tutorial 5

The thermochronological dataset presented in Table 5.7 was collected at a single location in an active mountain belt. Independent evidence suggests that the current tectonic setting has not changed since its inception. The data suggest a minimum value for the time of initiation of the current exhumation phase. What is it? Can we also derive from this limited thermochronological dataset a mean exhumation rate and a value for the pre-orogenic geothermal gradient? Assume that $\kappa = 25 \, \text{km}^2 \, \text{Myr}^{-1}$.

Table 5.6. *A thermochronological dataset obtained from fission-track analysis on apatite and zircon and from K–Ar dating on biotite and muscovite; to be used in Tutorial 4*

Age (Myr)	Temperature (°C)
2.5	115
7.1	280
10.7	380
412.1	410

Table 5.7. *A thermochronological dataset obtained from fission-track analysis on apatite and zircon and from K–Ar dating on biotite; to be used in Tutorial 5*

Age (Myr)	Temperature (°C)
1.9	115
4.9	250
11.9	375

Tutorial 6

As an application of the general solution to the transient one-dimensional heat-transport equation, we look at the effect of an abrupt decrease in exhumation rate caused by cessation of tectonic activity (the end of an orogenic event). We assume that rocks were being exhumed at a rate $\dot{E}_0 = 2 \, \text{km} \, \text{Myr}^{-1}$ during a tectonic event that started 12 Myr ago. In the recent past, i.e. $t_0 = 5$ Myr ago, the exhumation rate started to decrease exponentially according to

$$\dot{E} = \dot{E}_0 e^{t-t_0} \tag{5.67}$$

where t is time in the past. Using Heat1D, compute the thermal history of a rock particle that has reached the surface today. Ignoring heat production, assume that the layer being actively exhumed is 25 km thick and that its basal temperature is 600 °C. Compute the ages recorded by the (U–Th)/He and fission-track chronometers in apatite and for the K–Ar chronometer in biotite and muscovite. What can you conclude from the apparent ages of the rock sample? Redo the same experiment, assuming that the end of the tectonic event took place 2 Myr ago.

6
Steady-state two-dimensional heat transport

Until here, we have considered only one-dimensional cases, treating the surface as if it were completely horizontal. However, when considering ages corresponding to low-temperature geochronometers, it is important to consider the effect of finite-amplitude surface topography on the temperature distribution within the crust. Even though surface topography is usually rather complex and three-dimensional, for the sake of clarity and simplicity, we consider first how a periodic (sinusoidal), two-dimensional topography affects the transport of heat through the isothermal (0°C) upper surface. We consider the steady-state case in which heat transport is dominated by conduction, then proceed to include the effect of heat advection by exhumation and, finally, we consider the general transient problem.

6.1 The effect of surface topography

Conductive equilibrium

Assuming that heat production is negligible and that conductivity is uniform, the two-dimensional, steady-state heat-transport equation can be derived from Equation (4.15) to give

$$k \frac{\partial^2 T}{\partial x^2} + k \frac{\partial^2 T}{\partial z^2} = 0 \tag{6.1}$$

The effect of a periodic surface topography along which temperature is held constant can be included through the following surface boundary condition:

$$T = 0 \quad \text{on } z = z_0 \cos\left(\frac{2\pi x}{\lambda}\right) \tag{6.2}$$

where z_0 and λ are the amplitude and wavelength of the assumed surface topography, respectively. This boundary condition is commonly approximated by assuming that, along an imaginary surface defined by $z = 0$, the temperature varies according to

$$T = G_S z_0 \cos\left(\frac{2\pi x}{\lambda}\right) \quad (6.3)$$

where G_S is the temperature gradient near the surface. Note that this approximation will render the solution rather inaccurate for large-amplitude surface topography, especially near the surface. We also assume that the temperature is held fixed at $T = T_L$ at a depth $z = L$ much greater than the wavelength and amplitude of the topography.

This equation has a simple solution (Turcotte and Schubert, 1982):

$$T(x, z) = T_L \frac{z}{L} + \Delta \cos\left(\frac{2\pi x}{\lambda}\right) \exp\left(-\frac{2\pi z}{\lambda}\right) \quad (6.4)$$

where, to first order,

$$\Delta = T_L \frac{z_0}{L} \quad (6.5)$$

From the form of this approximate solution, one can readily see that the perturbation brought about by the finite-amplitude topography is strongly dependent on the wavelength of the topography. In fact, the perturbation decays exponentially with depth at a rate that is inversely proportional to the wavelength of the topography; the amplitude of the topography also affects the amplitude of the disturbance, but only linearly. One usually refers to the depth of penetration of the temperature disturbance as the 'skin depth', i.e. the depth at which the perturbation has dropped by a factor $1/e$, which in this case is simply equal to the wavelength of the topography.

The solution is displayed in Figure 6.1, where the geometry of isotherms is shown beneath a surface topography of amplitude 2 km and wavelength 10 km (panel (a)), an amplitude 1 km and wavelength 10 km (panel (b)) and an amplitude 2 km and wavelength 5 km (panel (c)), clearly demonstrating the stronger dependence of the perturbation on the wavelength of the topography than on its amplitude.

Effects of exhumation

When exhumation plays an important role in transporting heat towards the free, upper surface, the steady-state equation to solve is

$$-\rho c \dot{E} \frac{\partial T}{\partial z} = k \frac{\partial^2 T}{\partial x^2} + k \frac{\partial^2 T}{\partial z^2} \quad (6.6)$$

6.1 The effect of surface topography

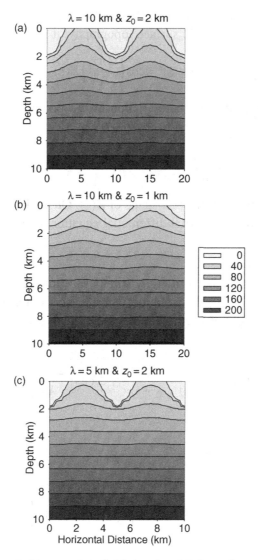

Fig. 6.1. A computed temperature field showing the dependences of the thermal perturbation caused by finite topography on wavelength and amplitude. Note that the horizontal scale is twice as large in panel (c) as in panels (a) and (b). The penetration depth of the disturbance is a strong function of the wavelength of the topography (compare panels (a) and (c)) in comparison with its amplitude (compare panels (a) and (b)). The assumed mean geothermal gradient, G_S, is $20\,°\mathrm{C\,km^{-1}}$.

or, in terms of the diffusivity, κ:

$$\dot{E}\frac{\partial T}{\partial z} + \kappa\frac{\partial^2 T}{\partial x^2} + \kappa\frac{\partial^2 T}{\partial z^2} = 0 \tag{6.7}$$

Assuming that the perturbation to the flat-Earth solution is in phase with (i.e. has the same shape as) the periodic surface topography, it may be written as

$$T'(x, z) = \cos\left(\frac{2\pi x}{\lambda}\right) Z(z) \tag{6.8}$$

where $Z(z)$ describes the vertical variation of the perturbation. By substitution into (6.7), one obtains

$$\kappa \frac{\partial^2 Z}{\partial z^2} + \dot{E} \frac{\partial Z}{\partial z} - \kappa \left(\frac{2\pi}{\lambda}\right)^2 Z = 0 \tag{6.9}$$

which can be used to determine the shape of $Z(z)$. On applying standard methods for solving a second-order partial differential equation, one obtains for $Z(z)$

$$Z(z) = c_1 e^{m_1 z} + c_2 e^{m_2 z} \tag{6.10}$$

This general solution can be verified by simply substituting it into the differential equation (6.9). Combining this solution with the horizontal part of the solution (i.e. $\cos(2\pi x/\lambda)$), the flat-Earth solution and the boundary conditions leads to (Mancktelow and Grasemann, 1997):

$$T(x, z) = T_L \frac{1 - e^{-z\dot{E}/\kappa}}{1 - e^{-L\dot{E}/\kappa}} + \Delta \cos\left(\frac{2\pi x}{\lambda}\right) e^{-mz} \tag{6.11}$$

where

$$\Delta = \frac{\dot{E} T_L}{\kappa(1 - e^{-\dot{E}L/\kappa})} z_0$$

$$m = \frac{1}{2} \left[\frac{\dot{E}}{\kappa} + \sqrt{\left(\frac{\dot{E}}{\kappa}\right)^2 + \left(\frac{4\pi}{\lambda}\right)^2} \right] \tag{6.12}$$

This solution is shown in Figure 6.2 for three values of the exhumation rate. The relationships illustrated show the following key features:

- finite exhumation leads to a higher temperature gradient near the surface; this is the result of heat transport by advection as described in Chapter 5;
- at any given depth, exhumation dampens the amplitude of the thermal disturbance caused by finite-amplitude topography; and
- at any given temperature, the disturbance increases with exhumation rate.

Indirectly, the last two features result from the advection of the temperature field, and thus the thermal disturbance caused by the finite-amplitude topography, towards the surface.

6.1 The effect of surface topography

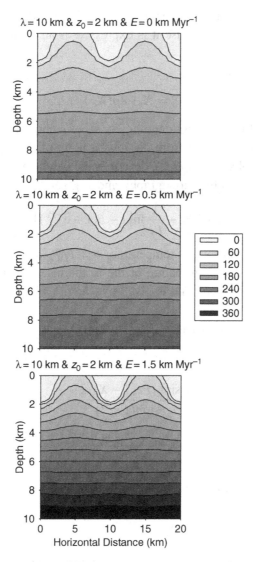

Fig. 6.2. Computed temperature fields showing the dependence of the thermal perturbation caused by finite topography on exhumation rate; the penetration depth of the disturbance is inversely proportional to the exhumation rate. The three panels correspond to $Pe = 0$, 0.5 and 1.5, respectively. Note that, in contrast to Figure 6.1, the horizontal and vertical scales are the same in all three panels.

Another approximate solution

The above solution (6.11) is not very accurate in the vicinity of the surface. Consequently, it cannot be used in practice to investigate the distribution of ages for low-temperature thermochronometers, since these are strongly influenced by the effects of the surface topography and the conductive transport of heat near the surface. Another approximate solution that is more accurate has been derived (Stüwe et al., 1994), but, because it is expressed as an infinite series, it is more difficult to compute.

The solution is given by

$$Z = X + \epsilon \rho(X, Y) \qquad (6.13)$$

in which Z is the dimensionless depth $Z = z/L$ of a chosen isotherm (where z is the depth positive downwards from the lowest point of the topography). ϵ is the scale of the topography, $\epsilon = z_0/L$. X describes the depth–temperature relationship in the eroding region without topography and is given by

$$X(\theta) = -\frac{1}{Pe}\ln\left[1 - \theta(1 - e^{-Pe})\right] \qquad (6.14)$$

in which θ is the dimensionless temperature $\theta = T/T_L$ and $Pe = \dot{E}L/\kappa$.

$\rho(X, Y)$ describes the perturbation of the isotherms calculated by use of X as a function of the topography and is given by

$$\rho(X, Y) = \alpha f(Y) + \sum_{n=1}^{\infty} \beta_n \int_0^{\infty} e^{-z}(f(Y - \gamma_n) - 2f(Y))dz \qquad (6.15)$$

in which

$$\alpha = \frac{\sinh[Pe(1-X)/2]}{\sinh(Pe/2)}$$

$$\beta_n = \frac{(-1)^{n+1} n\pi \sin[n\pi(1-X)]}{n^2\pi^2 + (Pe/2)^2} \qquad (6.16)$$

$$\gamma_n(z) = \frac{z}{\sqrt{n^2\pi^2 + (Pe/2)^2}}$$

and Y is the dimensionless lateral distance $Y = x/L$. $f(Y)$ is the normalised topography, which can be chosen to be of any shape but should be much smaller in vertical and lateral extent than L. The integral must be estimated numerically, for example by using a Gauss Laguerrian quadrature (Abramowitz and Stegun, 1970).

6.2 The age–elevation relationship – steady state

In tectonically active regions characterised by a finite-amplitude relief it is common practice to collect samples for thermochronological analysis along a 'vertical profile' or locally restricted range of elevations. Because they experience different exhumation paths and thus cooling histories, samples collected at the bottoms of valleys commonly record ages different from those of samples collected near the adjacent ridge tops. As we will show now, when performed properly, this sampling strategy is likely to provide good information on the exhumation rate.

If we consider the limited case of a thermochronological system that is characterised by a very high closure temperature, such that the finite-amplitude topography has only a negligible effect on the geometry of the corresponding isotherm, then one can readily see, as shown in Figure 6.3(a), that the slope of the relationship between age and elevation is a direct measure of the exhumation rate (Wagner and Reimer, 1972; Wagner et al., 1977; Fitzgerald and Gleadow, 1988; Fitzgerald et al., 1995). This is true if total exhumation has been sufficient to bring rocks to the surface both at the ridge tops and at valley bottoms that have crossed the closure-temperature isotherm during the same, current tectonic event, at a fixed exhumation rate, $\dot{E}$ (cf. Section 1.2). If the exhumation rate has changed during this time interval, it will result in a break in slope in the age–elevation relationship that can be used to date the time of change in the tectonic and/or erosional regime (cf. Section 1.2).

For a thermochronological system characterised by a low closure temperature, the corresponding isotherm is likely to be affected by the surface relief. If, as shown in Figure 6.3(b), the isotherm is disturbed by an amount αz_0, where z_0 is the amplitude of the surface topography, then the slope of the age–elevation relationship is $\dot{E}/(1-\alpha)$, and it provides an overestimate of the real exhumation rate. Because the steady-state (i.e. at thermal equilibrium) geometry of isotherms beneath a finite-amplitude periodic topography is approximately known (see Equation (6.11)), one can determine a correction factor with which to extract exhumation rates from age–elevation datasets for low-temperature systems. For fission-track ages (closure temperature of ~115 °C), the true exhumation rate $\dot{E}_T$ is related to the apparent exhumation rate (or slope of the age–elevation relationship), $\dot{E}_A$, by the following relationship:

$$\dot{E}_T = \dot{E}_A \frac{\Delta z}{z_0} \tag{6.17}$$

where Δz is the difference in depth to the closure-temperature isotherm beneath ridges and valleys. For the fission-track system, one can derive the following empirical relationship (Stüwe et al., 1994) between Δz and the exhumation rate:

$$\Delta z = a e^{-\dot{E}/b} \tag{6.18}$$

where a and b are constants that depend mildly on the wavelength and amplitude of relief (see Stüwe et al. (1994) for approximate values of and expressions for a and b).

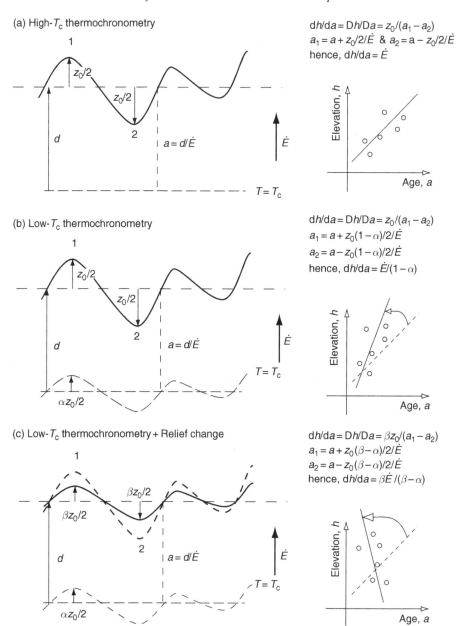

Fig. 6.3. Age–elevation relationships for (a) high-temperature systems, (b) low-temperature systems and (c) including the effect of a varying surface relief. After Braun (2002b). Reproduced with permission from Elsevier.

6.3 Relief change

During periods of active tectonics, rock uplift leads to vertical movements of the surface which, in turn, is reshaped by geomorphic processes. It is therefore likely that, while rocks are advected towards the surface by denudation, the shape of the surface is actively changing. Because most geomorphic processes are related to the movement of water and/or solid rock downslope, the positions of large-scale geomorphic features (valleys and ridges) are relatively stable. It is their scale, more precisely the amplitude of the relief, that is likely to change by the proportionately greatest amount through time. We therefore initially consider the thermal consequences of a change in relief amplitude only.

If we return to the situation described in Figure 6.3, where rocks are exhumed at a constant velocity towards the surface, we realise that changing surface relief amplitude, as depicted in Figure 6.3(c), will perturb the slope of the age–elevation relationship. Assuming that, between the time the rocks passed through the closure temperature and the time they are exhumed at the surface, the surface relief has changed by a factor β, one can easily show that the slope of the age–elevation relationship is given by

$$\frac{\partial a}{\partial z} = \frac{\beta \dot{E}}{\beta - \alpha} \qquad (6.19)$$

This relationship demonstrates how sensitive age–elevation relationships are to changes in surface relief. In situations where relief has been substantially reduced in the relatively recent past ($\beta < 1$), the slope of the age–elevation relationship becomes steeper with decreasing β and can even become negative (if $\beta < \alpha$). When relief increases ($\beta > 1$), the slope becomes shallower with increasing β and tends towards an asymptotic value (for $\beta \to \infty$) equal to the real exhumation rate, $\dot{E}$.

Conclusion

Age–elevation relationships contain important information on the tectonic and geomorphic history of a given region of the crust. It is clear, however, that caution must be used when one attempts to interpret these relationships. To derive accurate estimates of the mean exhumation rate from the slope of an age–elevation relationship, one must apply corrections to take into account the effect of the finite-amplitude topography on the underlying thermal structure and the effects caused by recent changes in relief amplitude. However, these corrections require relatively good knowledge of the evolution through time of the shape of the surface topography, which, in most cases, is poorly constrained.

Tutorial 7

Rocks are collected along a transect in a tectonically active area (mean exhumation rate $\dot{E} = 1.2 \, \text{mm yr}^{-1}$) characterised by a finite-amplitude relief (amplitude $z_0 = 1 \, \text{km}$ and wavelength $\lambda = 10 \, \text{km}$). The layer being exhumed is 35 km thick and the temperature at the base of the layer is 400 °C. Compute the thermal histories of three rocks collected (1) at the top of a hill, (2) at mean elevation and (3) at the bottom of a valley from a temperature of 200 °C to the surface. Use the approximate expression (6.11) to compute the temperature field. Using the program MadTrax, compute the apparent fission-track ages of the three samples as well as the apparent exhumation rate derived from the age–elevation relationship. What can you conclude? Redo the experiment with different wavelengths, $\lambda = 1$ and 30 km; as well as different amplitudes, $z_0 = 0.5$ and 2 km. What can you conclude?

7

General transient solution – the three-dimensional problem

If one wishes to predict cooling-age estimates following complex tectonic scenarios (potentially including discrete exhumation episodes) or in complex situations where a finite-amplitude surface topography is evolving through time, one cannot use analytical or semi-analytical solutions or be limited by the assumption of two-dimensionality. In this chapter, we present a numerical model by means of which to calculate the solution to the three-dimensional heat-transport equation that is based on a finite-element method. The code, named Pecube, is made available to the readers. After a short description of the method used in Pecube, we use it to solve a problem involving a complex, evolving surface topography and demonstrate how age datasets can be used to constrain the rate at which geomorphic processes evolve.

7.1 Pecube

We first recall the expression for the general three-dimensional heat-transport equation (Section 4.6):

$$\rho c \left(\frac{\partial T}{\partial t} - \dot{E} \frac{\partial T}{\partial z} \right) = \frac{\partial}{\partial x} k \frac{\partial T}{\partial x} + \frac{\partial}{\partial y} k \frac{\partial T}{\partial y} + \frac{\partial}{\partial z} k \frac{\partial T}{\partial z} + \rho H \tag{7.1}$$

with the following general boundary conditions:

$$T(t, x, y, z = S(x, y, t)) = T_S(z)$$
$$= T_{\text{msl}} + \beta_{\text{r}}(z - z_{\text{msl}})$$
$$T(t, x, y, z = L) = T_L \tag{7.2}$$
$$\frac{\partial T}{\partial n} = 0 \text{ along the side boundaries}$$

where S describes the geometry of the surface, β_r is the atmospheric temperature lapse rate and z_{msl} and T_{msl} are fixed reference height and temperature (at mean sea level, for example). An initial temperature field is also defined:

$$T(t=0, x, y, z) = T_0(x, y, z) \tag{7.3}$$

Note that, in Equation (7.1), we have included only the vertical advection term, neglecting the potential thermal perturbation caused by lateral heat advection. This is because lateral temperature gradients are usually much smaller than vertical ones. In some tectonic situations, lateral heat advection may be important; we discuss this point in detail in Chapter 10.

Following the steps described in Section 5.3, we have recently developed a Fortran code called Pecube (Braun, 2003) that, although rather conventional in its conception, includes a novel approach to the incorporation of a time-varying surface geometry.

7.2 Time-varying surface topography

The finite-element grid in Pecube is made of vertical triangular or rectangular prisms (Figure 7.1) that permit any complex surface geometry to be connected to a flat base (where the fixed-temperature basal boundary condition is imposed). This geometry implies that nodes are aligned in vertical columns.

To account for the time-varying geometry of the upper surface, a Lagrangian approach is used, which consists of vertically translating the top nodes of the mesh by the required amount, $\dot{S}(x, y, t)\Delta t$, at the start of each time step. This method is accurate provided that $\Delta t < \Delta z/(10|\dot{S}|)$, where $\dot{S}$ is the rate of change of the surface topography. The vertical translation of the top nodes may lead to mesh distortion and an inaccurate solution to the equation. To circumvent this problem, the temperature field is interpolated from the deformed mesh onto an 'undeformed' mesh at the end of each time step. Because of the nature of

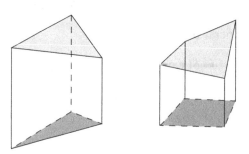

Fig. 7.1. Two types of three-dimensional prismatic elements used in Pecube.

7.2 Time-varying surface topography

the finite-element mesh used in Pecube this interpolation needs to be performed only in the vertical direction, thus minimising potential smoothing errors. The procedure is illustrated in Figure 7.2.

To improve stability in cases in which heat advection dominates over diffusion, a streamline-upwind Petrov–Galerkin method is used (Hughes and Brooks, 1982) and the shape-function matrix, N, in the advection term in the finite-element equation,

$$\int_{V_e} N^T \dot{E} \mathbf{B} \, dV_e \tag{7.4}$$

is replaced by

$$N^* = N + \tau \dot{E} B_3 \tag{7.5}$$

where

$$\tau = \frac{\Delta z \sqrt{15}}{|\dot{E}|} \tag{7.6}$$

and Δz is a vertical length scale (the thickness of the element).

A complete description of the program Pecube is given in Appendix 8.

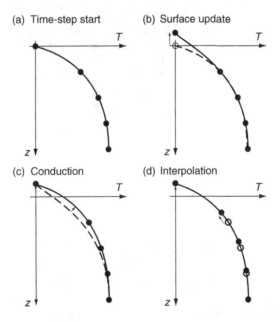

Fig. 7.2. In (a) and (b) we illustrate the Lagrangian method used to incorporate the effects of vertical movement of the upper boundary; (c) shows thermal relaxation over time step Δt; and (d) shows interpolation from the 'deformed' grid, represented by the open circles, onto a new 'undeformed' grid represented by the black circles.

7.3 Surface relief in the Sierra Nevada

To illustrate how sensitive thermochronological ages are to the evolution of surface relief with time, we have used Pecube to compute the distribution of ages at the Earth's surface predicted under two end-member scenarios for the geomorphic evolution of a high-relief area in the Sierra Nevada of western North America. In recent years, several apatite (U–Th)/He age datasets have been collected to constrain the exhumation and relief history of this area (House *et al.*, 1997, 1998, 2001). They are shown in Figure 7.3 as a simple age versus elevation plot.

There are several hypotheses regarding the origin of the present-day relief in the area (House *et al.*, 1998). The mooted contributing factors are rapid uplift and erosion during the Laramide Orogeny, some 70–100 Myr ago, and a more recent (late-Tertiary) episode of relief rejuvenation following local uplift and/or climate change. We have used Pecube to reproduce these two scenarios and test whether predicted ages differ significantly between them. Both scenarios assume that, at the end of the Laramide Orogeny some 70 Myr ago, the surface relief was approximately twice as large as today's. In the first scenario, relief amplitude is assumed to have decayed rapidly after the Laramide Orogeny (i.e. within 20 Myr)

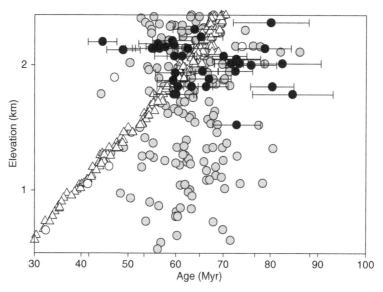

Fig. 7.3. Apatite (U–Th)/He ages collected in the Sierra Nevada region of western North America. The black circles correspond to the ages collected at an approximately constant elevation of 2000 m by House *et al.* (1998); the white circles are the ages collected by House *et al.* (1997) along a relatively steep and narrow profile along the side of Kings Canyon, a prominent geomorphic feature in the area. The grey circles and white triangles correspond to the Pecube predictions obtained assuming an 'old' decaying relief and a 'young' relief, respectively.

7.3 Surface relief in the Sierra Nevada

to one-tenth of its present-day value and to have been rejuvenated during the last 5 Myr; in the second scenario, relief amplitude is assumed to have decayed steadily over the last 70 Myr. In both cases, it is assumed that the temperature is fixed at 500 °C at the base of the crust ($z = 35$ km) and at 15 °C along the top surface (the lapse rate is neglected), and that the exhumation rate was high (1 km Myr^{-1}) during the Laramide Orogeny (between 100 and 70 Myr ago) and very low (0.03 km Myr^{-1}) between 70 Myr ago and the present. This slow mean exhumation of the area is likely to represent the isostatic rebound associated with the erosion of the mountain belt during its post-orogenic phase. The thermal diffusivity is set at 25 km^2 Myr^{-1}, and heat production is neglected. The problem is solved on a $51 \times 51 \times 35$-node mesh with a spatial resolution of 1 km in all directions. The geometry of the surface topography is extracted from a 1-km-resolution digital elevation model (DEM) (GTOPO30). Changes in relief are incorporated by modifying the amplitude of the topography, not its shape. This implies that the geometry of the drainage system (i.e. the location of the major river valleys) has not changed during the last 110 Myr.

The results are shown in Figure 7.4 as three-dimensional perspective plots of the finite-element mesh on the sides of which contours of the temperature field have been superimposed. Three critical times are shown: at the end the Laramide Orogeny and, for each experiment, 20 Myr later and at the end of computations (i.e. the present day). The contours of temperature clearly show the effect of vertical heat advection, especially at the end of the orogenic phase (Figure 7.4(a)), when the isotherms are compressed towards the surface and deformed by the high-relief surface topography. After 20 Myr, the two solutions are different: under scenario 1 (Figure 7.4(b)), the system has almost reached conductive equilibrium beneath a flat surface, whereas, under scenario 2 (Figure 7.4(c)), the low-temperature isotherms are deformed by the presence of a high-relief topography. The predicted present-day temperature structures are relatively similar in both scenarios, except that, in the first case (Figure 7.4(d)), the topography is too young to affect the underlying thermal structure, whereas in the second case (Figure 7.4(e)), the isotherms are perturbed by the finite-amplitude surface topography.

Pecube predicts $T-t$ paths for all rock particles that, at the end of computations, occupy the locations of the nodes along the top surface of the finite-element mesh. From these $T-t$ paths, an apparent (U–Th)/He age for apatite can be computed at each location, following the procedure and parameter values described in Section 2.5. Colour contours of the predicted ages have been superimposed on the surface topography of the last two panels of Figure 7.4. The computed mean ages are relatively similar (60.94 and 68.71 Myr, respectively). These depend mostly on the assumed age for the end of the Laramide Orogeny. The distributions of ages on the landscape are, however, very different (compare Figures 7.4(d)

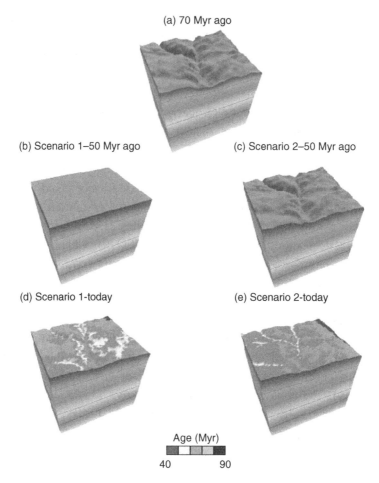

Fig. 7.4. A perspective view of the finite-element domain with the temperature field colour-contoured on the sides. (a) The solution at the end of the Laramide Orogeny; (b) 20 Myr later following scenario 1 and (c) 20 Myr later following scenario 2; (d) the solution at the end of the computations (corresponding to the present day) obtained following scenario 1; and (e) the corresponding solution obtained following scenario 2. Age contours are superimposed on the surface in panels (d) and (e).

and (e)). Following the first scenario (Figure 7.4(d)), most ages range between 30 and 70 Myr with a very clearly defined linear relationship between age and elevation. In the second scenario (Figure 7.4(e)), the range of predicted ages is similar (40–90 Myr) but their relationship to elevation is less clear. On the scale of a single river valley (<10 km), ages are proportional to elevation but, on the larger scale (>10 km), older ages are found at lower elevations, i.e. near the valleys.

7.3 Surface relief in the Sierra Nevada

Thus, while the two scenarios lead to very similar predictions for the present-day thermal structure beneath Kings Canyon and the mean (U–Th)/He ages in apatite, they predict very different relationships between age and elevation. This is further documented in Figure 7.3 where the predicted ages are plotted against surface elevation for the two scenarios and compared with the data collected by House et al. (1997, 1998). In the first scenario (white triangles), there is a clear linear relationship between age and elevation. The slope of the regression line between age and elevation is related to the exhumation rate (Stüwe et al., 1994). In the second scenario (grey circles), this relationship is not so clear and ages vary by as much as 50 Myr at any given elevation. The first dataset (black circles), collected across a limited range of elevations (1.8–2.2 km) exhibits a pattern similar to the predictions of scenario 2 (i.e. large age variations at constant elevation). However, the second dataset (white circles), collected along the valley wall of Kings Canyon (House et al., 1997), across a very short distance (<2.5 km) but covering a larger range of elevations (0.5–2.2 km), exhibits a clear linear relationship between age and elevation, which is much more similar to the predictions derived from scenario 1.

This apparently contradictory result is easily explained by considering the influence of the wavelength of topography on the thermal perturbation it causes and its consequences for thermochronological ages. When collected across a steep valley wall, an age–elevation dataset provides constraints on the mean exhumation rate of rocks in the area, not its geomorphological evolution (cf. Section 6.2). A thermochronological dataset must be collected across broad features of the landscape if it is to contain information about landform evolution (cf. Section 6.3). Thus the first of the two databases (black circles in Figure 7.3) is suitable to constrain the evolution of relief and its comparison with the model results suggests that the second scenario (grey circles in Figure 7.3) is correct, implying a slow reduction in topographic relief over the past 50–60 Myr. This point is further developed in Section 8.1 and illustrated in Section 8.3.

Tutorial 8

Using Pecube, compute age–elevation relationships for the (U–Th)/He apatite and K–Ar muscovite thermochronometers. Consider three scenarios in which a 5-Myr-old tectonic event leads to exhumation at 2.5 km Myr^{-1}. In the first scenario, assume that the relief has increased from 0.3 times its present-day value over the last 100 000 yr. In the second scenario, the relief has remained unchanged. In the third scenario, the relief has decreased from 1.8 times its present-day value over the last 100 000 yr. Extract the surface topography from the default dataset (Topo.dat). What can you deduce from the age–elevation relationships?

8
Inverse methods

So far, we have provided the reader with a variety of mathematical methods, including analytical, semi-analytical and numerical ones, with which to solve the heat-transport equation in a wide range of situations. The purpose of these calculations has been to produce thermal histories that are then used to predict thermochronological ages of a form that allows their comparison with real data that a geoscientist might be able to acquire. Through this comparison, we hope to provide constraints on tectonic or geomorphic processes.

In this chapter, we propose a different approach in which the thermochronological data are directly 'inverted' to provide quantitative estimates of parameters that are direct measures of a tectonic or geomorphic process, such as the mean rate of rock exhumation (which is related to tectonic uplift) and the rate of change of surface relief. As we will show in later chapters, these 'inverse' methods can also be used to provide direct constraints on a wide range of parameters characterising tectonic events, such as the geometry of active faults, the thermal properties of the crust and the evolution of surface relief through time.

8.1 Spectral analysis

In Section 6.1, we showed how the thermal perturbation caused by finite-amplitude surface topography decreases exponentially with depth at a rate that is proportional to the wavelength of the topography (Turcotte and Schubert, 1982; Stüwe et al., 1994; Mancktelow and Grasemann, 1997). As illustrated in Figure 8.1, for a given isotopic system, there exists a critical wavelength (λ_c) below which topography has little effect on the shape of the corresponding closure/blocking-temperature isotherm. Thus, the slope of the relationship between age and elevation data collected on a scale smaller than λ_c should provide an accurate estimate of the exhumation rate (Figure 8.1(a)). Conversely, at wavelengths much larger than λ_c,

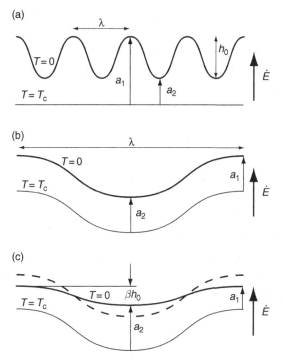

Fig. 8.1. (a) Short-wavelength topography does not affect the geometry of the closure-temperature ($T = T_c$) isotherm and the slope of the age–elevation relationship is the inverse of the exhumation rate, $\dot{E}$,

$$\frac{\partial a}{\partial h} = \frac{a_1 - a_2}{h_0} = \frac{1}{\dot{E}}$$

(b) Long-wavelength topography strongly affects the shape of the T_c isotherm and age is independent of elevation ($\partial a / \partial h = 0$) (c) At long topographic wavelengths, any age variation with elevation is indicative of a relative change in relief amplitude since rocks crossed the T_c isotherm. After Braun (2002a). Reproduced with permission from Blackwell Publishing.

the closure-temperature isotherm follows exactly the shape of the topography and there should be no variation in age with elevation (Figure 8.1(b)), unless the relief has changed since the rocks passed through the closure-temperature isotherm (Figure 8.1(c)).

Braun (2002a) proposed a method that makes use of the fractal nature of surface topography (Huang and Turcotte, 1989) to sample the relationship between age and elevation for a wide range of topographic wavelengths by collecting data along one-dimensional transects. Extracting the relationship between age and elevation is equivalent to determining the so-called 'frequency-response function' or 'admittance function' of a system that has elevation as input and age measurements derived from rocks sampled along the profile as output (Jenkins and Watts, 1968). The response function can be described in terms of a gain, G, and a phase, F. Both are functions of the wavelength of the input topography, λ. Expressions for $G(\lambda)$ and $F(\lambda)$ can be obtained from classical spectral analysis (Jenkins and Watts, 1968):

$$G(\lambda) = \frac{\sqrt{C_{12}^2 + Q_{12}^2}}{C_{11}}$$

$$F(\lambda) = \arctan\left(-\frac{Q_{12}}{C_{12}}\right)$$

(8.1)

where C_{12} and Q_{12} are the real and imaginary parts of the cross spectrum obtained from the real and imaginary parts of the smoothed spectral estimators of the input and output signals. These estimators are, in turn, obtained from the Fourier transforms of the windowed input (elevation, z) and output (age, a) signals, R_z, I_z and R_a, I_a (Jenkins and Watts, 1968):

$$C_{12} = R_z R_a + I_z I_a$$
$$Q_{12} = I_z R_a - R_z I_a$$

(8.2)

C_{11} is the power spectrum of the input signal:

$$C_{11} = R_z^2 + I_z^2$$

(8.3)

The smoothed spectral estimators are obtained by applying a non-rectangular window (Jenkins and Watts, 1968) to the elevation and age profiles prior to the calculation of the Fourier transforms. This windowing is necessary in order to obtain statistically meaningful estimates of the spectral information (Jenkins and Watts, 1968).

8.2 An example based on synthetic ages

To illustrate how the method can be used to interpret thermochronological data, a 128-km-long elevation profile has arbitrarily been extracted from a 1-km-resolution DEM (GTOPO30) of South Island, New Zealand, in a direction parallel to the Alpine Fault, the main crustal-scale structure accommodating the oblique convergence between the Pacific and Australian plates (Batt et al., 2000). Hence the input (elevation) signal has a spectral content that is representative of a natural landform. Moreover, being parallel to the Alpine Fault and only a few tens of kilometres in length, the transect should be characterised by a spatially uniform exhumation rate, $\dot{E}$. The elevation profile is shown in Figure 8.2(a). The Pecube

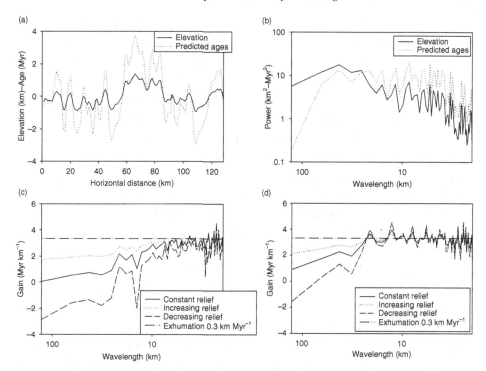

Fig. 8.2. (a) An elevation profile extracted from a 1-km DEM of South Island (New Zealand) and predicted (U–Th)/He ages. The mean value has been subtracted from both profiles. (b) Power spectra of the elevation and age profiles. (c) The real part of the gain function between age and elevation calculated from (8.1) assuming three different scenarios for relief evolution. (d) The real part of the gain function for ages corresponding to a closure temperature of 300 °C. Because only vertical tectonic transport is considered here, at all wavelengths, variations in age must be in phase or exactly out of phase with variations in elevations, i.e. the phase estimates are either 0 or $\pm\pi$. Thus, only the real part of the gain is shown. After Braun (2002a). Reproduced with permission from Blackwell Publishing.

code (see Chapter 7) was used to solve the heat-transport equation in three dimensions and to compute the temperature field that was subsequently used to derive temperature–time paths. From these T–t paths, an age profile for the (U–Th)/He chronometer was derived using the subroutine Mad_He (see Section 2.5). For the sake of illustrating the method, a uniform exhumation rate of $0.3 \, \text{km} \, \text{Myr}^{-1}$ and a conductive geothermal gradient of $10 \, °\text{C} \, \text{km}^{-1}$ were arbitrarily assumed.† The thermal diffusivity was set at $25 \, \text{km}^2 \, \text{Myr}^{-1}$ and radioactive heat production was neglected. The computed age profile is shown in Figure 8.2(a). The calculated

† Note that these values of the exhumation rate and conductive geothermal gradient were chosen to illustrate the method and should not be taken as factual values for any particular region of the Southern Alps of New Zealand, which, we know, are characterised by much faster mean exhumation rates (Batt et al., 2000).

power spectra are shown in Figure 8.2(b) and indicate how the amplitude of the elevation and age signals is distributed at various horizontal length scales. The computed gain is shown in Figure 8.2(c) and shows that, at wavelengths shorter than 8 km, the gain ($\sim 3\,\text{Myr}\,\text{km}^{-1}$) provides a good estimate of the inverse of the imposed exhumation rate ($0.3\,\text{km}\,\text{Myr}^{-1}$). This arises because isotherms are not perturbed by topography at short wavelengths (see Figure 8.1(a)). At intermediate wavelengths, the gain is smaller than the inverse of the imposed uplift rate as the finite topography starts to perturb the closure-temperature isotherm for (U–Th)/He in apatite. At wavelengths larger than 25 km, the gain tends towards zero as the closure-temperature isotherm becomes parallel to the surface topography and ages therefore become independent of elevation (see Figure 8.1(b)).

Figure 8.2(c) also shows, as a short-dashed line, the gain derived from a model experiment in which the surface relief (the amplitude of the surface topography) is progressively increased to its present value from a previous condition at half that level over the last 10 Myr of the numerical experiment. At short wavelengths, the results are similar to those of the first experiment, but, at large wavelengths, the gain estimates tend asymptotically towards a finite, positive value of $\sim 1.8\,\text{Myr}\,\text{km}^{-1}$. Similarly, in an experiment in which surface relief is decreased from twice its present-day value over the last 10 Myr, the predicted gain values (long-dashed line in Figure 8.2(c)) tend towards a finite negative value of $\sim -2.5\,\text{Myr}\,\text{km}^{-1}$ at long wavelengths.

The computed admittance/gain function therefore contains two independent pieces of information: (1) the asymptotic value of the gain at short wavelengths, G_S, provides a direct estimate of the inverse of the mean exhumation rate; and (2) the asymptotic value of the gain estimate at long wavelengths, G_L, indicates whether, in the recent past, the topography has remained constant ($G_\text{L} = 0$), increased ($G_\text{L} > 0$) or decreased ($G_\text{L} < 0$) in amplitude. As shown in Figure 8.1(c), at very long wavelengths, any gradient in age with elevation must be related to recent changes in surface relief, i.e. changes experienced since the rocks passed through the closure-temperature isotherm. If β is defined as the ratio between relief amplitude now and at the time rocks passed through the closure temperature, the gradient in age with elevation, $\partial a/\partial z$, is equal to (Figure 8.1)

$$\left(\frac{\partial a}{\partial z}\right)_\text{L} = \frac{\beta - 1}{\beta}\frac{1}{\dot{E}} \tag{8.4}$$

and one can write, since $G_\text{S} = 1/\dot{E}$,

$$\beta = \frac{1}{1 - G_\text{L}/G_\text{S}} \tag{8.5}$$

The ratio G_L/G_S provides, therefore, a direct estimate of the relative change in surface relief over the time period defined by the mean thermochronological age recorded. This result is based on the assumption that, when the rocks pass through the closure-temperature isotherm, the temperature field is near steady state (i.e. the isotherms are in phase with the long-wavelength topography); if this is not the case, the ratio G_L/G_S must be regarded as a maximum estimate of the relative change in relief. The values of the relief-reduction factor, β, computed from the gain values shown in Figures 8.2(c) are 0, 2.5 and 0.55, which compare very well with the imposed values of 0, 2 and 0.5, respectively.

8.3 Application of the spectral method to the Sierra Nevada

One of the datasets collected in the Sierra Nevada (House et al., 1997, 1998, 2001), which we briefly described in Section 7.3, extends for approximately 180 km across the southern Sierra Nevada batholith and contains 36 apatite (U–Th)/He age determinations from rocks collected at or near an elevation of 2000 m (House et al., 1998). This sampling strategy is not ideally suited to our method of analysis but no better dataset exists at present. The data were interpolated to provide 128 equally spaced model ages that were used to compute the spectra and gain. The results are shown in Figure 8.3.

At short wavelengths, the gain values are not very constrained. This is because the data sampling is too coarse and irregular to capture the relationship between age and elevation at wavelengths below 10 km. An independent dataset collected along a very short transect (<2.5 km) in the Kings Canyon area does exist, however (House et al., 1997). A linear regression between (U–Th)/He age and elevation from this dataset yields an exhumation rate of $0.04\,\text{km}\,\text{Myr}^{-1}$. The inverse of this value ($25\,\text{Myr}\,\text{km}^{-1}$) is shown in Figure 8.3(c) as a horizontal dashed line. Although the plot is very noisy, the gain values at short wavelengths are consistent with this estimate.

At long wavelengths, the real part of the gain is clearly negative, indicating that relief has decreased during the last 70 Myr. The asymptotic value of the gain at very long wavelengths is approximately $-25\,\text{Myr}\,\text{km}^{-1}$, which, combined with the estimated exhumation rate of $0.04\,\text{km}\,\text{Myr}^{-1}$ and using (8.5), yields a value of $\beta = 0.5$ for the predicted relative reduction in relief. This estimate agrees well with that obtained by House et al. (1998) by interpreting age variations at constant elevation as recording the thermal structure related to paleotopography (e.g., Mancktelow and Grasemann, 1997) (cf. Section 6.1). It supports their suggestion that, since the end of the Laramide Orogeny some 70–80 Myr ago, the surface relief in the Sierra Nevada has decreased by more than 50%. In contrast to the estimate of relative relief derived by House and co-workers, that calculated here

128 *Inverse methods*

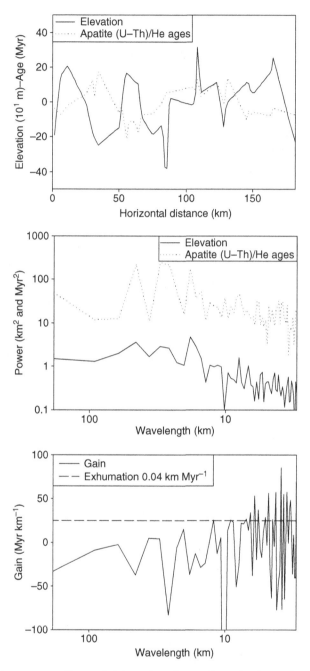

Fig. 8.3. The same as Figure 8.2 but elevation and apatite (U–Th)/He age data are from a transect through the Sierra Nevada (House *et al.*, 1998). Mean values of 2 km and 67 Myr have been subtracted from the elevation and age datasets, respectively. After Braun (2002a). Reproduced with permission from Blackwell Publishing.

from the admittance function does not rely on any assumption on the value of the past or present geothermal gradient. This example demonstrates the power of spectral analysis as an inverse method to derive accurate and independent estimates of denudation rates and rates of change of surface topography from age–elevation datasets.

8.4 Sampling strategy

To extract optimal benefit, the admittance method described here requires a specific sampling strategy that ensures that elevation and age are measured at the appropriate range of wavelengths. The optimal sampling interval is approximately half an order of magnitude smaller than λ_c, the critical wavelength introduced earlier, and the optimal length of the profile is half an order of magnitude larger than λ_c. The dataset should therefore contain a minimum of 30–50 samples. Because it depends on the closure temperature, λ_c is different for each thermochronometric system. It also depends on the geothermal gradient and the mean exhumation rate (see Section 6.1). Attention must therefore be paid to these factors when designing a sampling strategy, with wider spacing between samples allowed for high-temperature thermochronometers and/or cold, slowly eroding tectonic environments. This is illustrated in Figure 8.2(d), where the gain values are derived from ages calculated for a higher-blocking-temperature geochronometer such as K–Ar in biotite ($T_c = 300\,°C$). The transition from high to low gain values takes place at a longer wavelength.

The spectral method also demonstrates that, although thermochronological datasets contain information on landform evolution, this information is limited to the features of the landscape that are larger than λ_c. This implies that, whatever the method used to interpret it, no information on the rate of evolution of the landscape can be extracted from a thermochronological dataset that has been collected across a small-scale feature of that landscape.

It is also important to stress that the spectral method requires that both datasets be stationary, i.e. the way in which age is related to elevation must be similar along the length of the transect. Thus, in choosing the location of the transect, one must ensure that the geothermal gradient, exhumation rate and relief characteristics are uniform across the area being studied. It is also clear that the method can provide estimates of the exhumation rate and rate of change in surface relief only over a time interval that corresponds to the mean age of the samples. As suggested in Batt and Braun (1997), the use of several thermochronometers is therefore required when one wants to invert age–elevation datasets obtained in areas affected by a complex, multi-stage tectonic or geomorphic history.

8.5 Systematic searches

In previous sections, we have developed an array of analytical and numerical methods to predict temperature histories from given tectonic and geomorphic scenarios, which can be used to produce apparent thermochronometric ages. These synthetic ages are compared with existing data in order to estimate the validity of the chosen scenarios. This process can be time-consuming, especially if the scenarios are complex, i.e. if they depend on a large number of relatively unconstrained parameters.

Inverse methods have been developed to automate this process. Most inverse methods (Press *et al.*, 1986) rely on a directed search through the parameter space to determine the set of parameter values that, through the forward model, predict the best fit to the available data. Many inverse methods have been devised; the most successful ones are capable of finding the best global fit to the data, i.e. the one that minimises the misfit between data and model predictions. In highly non-linear problems, this search may be impaired by the existence of so-called 'local minima', which produce a relatively low misfit but not the global minimum. Owing to the quasi-linear nature of the heat-transport equation, predicting thermochronometric ages from tectonic scenarios is a relatively linear process and standard inverse methods (simple Monte Carlo searches) usually work well. When complex tectonic scenarios are envisaged, i.e. those that involve a relatively large number of free parameters (five or more), more sophisticated methods must be used, such as the Neighbourhood Algorithm (NA) (Sambridge, 1999a, 1999b).

We refer the reader to recent works in which thermochronological data have been inverted by Monte Carlo simulation (Gallagher, 1997; Willett, 1997; Willett, *et al.*, 1997; Moore and England, 2001). Examples of using data-inversion techniques (and especially NA) to constrain evolutionary scenarios are provided in Chapters 11, 12 and 13.

9

Detrital thermochronology

A relatively recent development in thermochronology is the analysis of samples from the erosional products of mountain belts, which is known as detrital thermochronology. The main advantage of detrital thermochronology over the more classical analysis of in situ samples is that the evolution of cooling and denudation rates through time can be monitored by analysing samples from sediments of different ages. Thus, detrital thermochronology provides a much longer 'memory' of denudation rates than does in situ thermochronology, especially in regions of high denudation rates and therefore young thermochronological ages. The gain in temporal range is, however, counterbalanced by a reduction in spatial resolution: it is not always clear what the source areas of the detrital samples were. In this chapter, we will discuss the basic approach of the detrital method and the quantitative interpretation of detrital thermochronological data in terms of the temporal evolution of source-area denudation rates and/or relief. We also consider the influence of post-depositional partial resetting on the interpretation of detrital data.

9.1 The basic approach

The long-term evolution of mountain belts is closely controlled by the interplay among tectonics, surface erosion and deposition (as discussed in detail in Chapter 13). Thermochronology provides us with insights into the exhumation history of rocks within the orogen and the use of multiple thermochronometers, or of multiple vertically separated samples, may help us to elucidate the evolution of the mountain-belt system within a thermal or spatial reference frame, respectively (cf. Section 1.2). However, samples collected from bedrock exposed at the surface within the mountain belt ('*in situ*' samples) will allow constraint of the thermal history only between open-system and blocked behaviour relevant to the specific thermochronological system studied. They therefore have only limited 'memory', information about previous thermo-tectonic events having been removed by the erosion of the overlying rocks that contained that record. Thus, in active and rapidly denuding orogenic settings such as the Southern Alps of New Zealand, Taiwan and the Himalayas, *in situ* bedrock samples may inform us only about the thermal evolution during the last few million (for high-temperature systems) or even few hundred-thousand (for low-temperature systems) years (e.g., Tippett and Kamp, 1993; Batt *et al.*, 2000; Willett *et al.*, 2003; Burbank *et al.*, 2003; Thiede *et al.*, 2004).

In contrast to the relatively short-term information accessible from bedrock surface samples, orogenic material contained in the sedimentary basins that generally surround mountain belts contains a much longer-term record of the evolution of their source areas. Sedimentary geologists have, for the past several decades, invested significant effort in extracting information about sediment provenance and the tectonic and climatic history of source areas from the stratigraphy and sedimentology of foreland basin sediments (e.g., Burbank, 1992; Schlunegger, 1999), as well as from their mineralogical, chemical and isotopic characteristics (e.g., Harrison *et al.*, 1993; Garzanti *et al.*, 1996; Huyghe *et al.*, 2001). Thermochronological dating techniques have recently been applied to detrital grains recovered from orogenic sediments, in order to track the long-term thermal and exhumational histories of their source areas. From its initial development with the publication of detrital fission-track data from the Himalayan foreland basin deposits in Pakistan by Cerveny *et al.* (1988), this approach has rapidly been expanded to other chronometers and settings. Copeland and Harrison (1990) and Harrison *et al.* (1993) produced $^{40}Ar/^{39}Ar$ data for detrital micas from Himalayan sediments in the Bengal Fan and Nepal, respectively, whereas Brandon and Vance (1992) provided the first methodological treatment for detrital zircon fission-track thermochronology. Subsequent methodological reviews have been provided by Lonergan and Johnson (1998), Garver *et al.* (1999), von Eynatten and Wijbrans (2003) and Bernet *et al.* (2004), and the technique is today widely accepted as an important element in the quantitative constraint of orogenic evolution.

Central to the interpretation of detrital thermochronological ages is the concept of 'lag time', that is, the difference between the stratigraphic age of the sediment from which a sample was taken and that sample's thermochronologic or cooling age (Figure 9.1). This time difference is generally assumed to represent the time taken for the sample to be exhumed from its closure depth to the surface; the transport time between exposure at the surface and deposition in the sedimentary basin is thus considered negligible (Brandon and Vance, 1992). It follows that the lag time contains information on the average exhumation rate within the orogen leading up to the time of deposition of the sample; variations in denudation rate

9.1 The basic approach

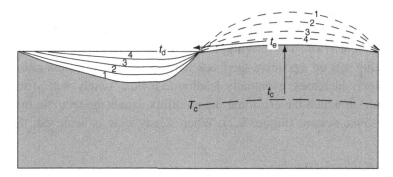

Fig. 9.1. Principles of detrital thermochronology and the concept of 'lag time'. The source area is progressively unroofed and the erosional products are deposited in the adjacent sedimentary basin. Arrows indicate highly simplified particle paths, consisting of exhumation towards the surface followed by transport towards the basin. The lag time is the age difference between the depositional age (t_d) of the sediment from which the sample was taken and the sample's cooling age (t_c), i.e., the time at which it passed through its closure isotherm (T_c). The time of deposition (t_d) is supposed to be the same as the time of exposure at the surface (t_e), i.e., the transport time is negligible. Modified from Garver *et al.* (1999). Reproduced with permission by the Geological Society of London.

through time will be recorded by variations in lag time upsection. The resolution to which the depositional age can be constrained will limit the precision of the analysis. Samples are therefore preferentially collected from bio-stratigraphically or (more generally) magneto-stratigraphically dated sections.

The thermochronological age of a detrital sample might not be very well defined; detrital single-grain ages usually exhibit considerable spread, so the use of a mean or pooled age is meaningless. Techniques to deconvolve the single-grain ages of sedimentary samples into statistically meaningful age groups have been developed (cf. Section 9.2). The thermochronological age considered in the calculation of a lag time is then one of the population age peaks (usually the youngest). Lag times can also be calculated for each single-grain age determined and their distribution interpreted in terms of spatial variations in source-area denudation rates or relief (cf. Sections 9.3 and 9.4).

Systematic variations in lag-time distribution (and therefore denudation rate of the source area) are interpreted as pertaining to different stages in the orogen's evolution. This relationship can be illustrated with reference to the simple kinematic model of Jamieson and Beaumont (1989), in which the coupled tectonic and erosional evolution of an orogenic system are quantitatively simulated in two dimensions (see also Beaumont *et al.* (1999); Willett *et al.* (2001) and Willett and Brandon (2002)). As orogenesis gets under way, the tectonic mass influx into the orogenic system will be greater than the erosional outflux, leading to crustal thickening, surface uplift and the creation of topography during the 'constructional' phase of a collisional orogen. Because the rate of erosion by both hillslope processes and fluvial processes is strongly controlled by slope (e.g., Beaumont et al., 1999; Whipple and Tucker, 1999), the creation of topographic relief sets up a positive-feedback loop in which denudation rates increase as topography increases, eventually leading to a flux 'steady state' (or dynamic equilibrium) in which the erosional mass outflux equals the tectonic mass influx in the orogenic system (Figure 9.2). When steady state is achieved, the orogen

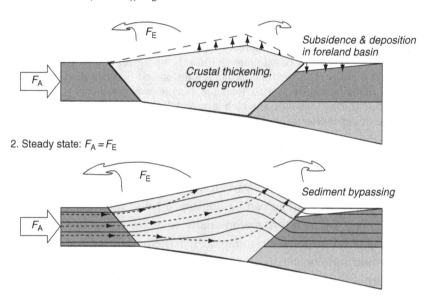

Fig. 9.2. Evolutionary stages in a collisional mountain belt (after Jamieson and Beaumont (1989); Willett and Brandon (2002)) and their detrital thermochronologic record. During the constructional phase (1), the tectonic mass influx (F_A) is greater than the erosional mass outflux (F_E), leading to crustal thickening, orogen growth and an increase in denudation rates with time. At steady state (2), the tectonic influx equals the erosional outflux, the orogen ceases to grow and denudation rates become constant with time. During the destructional phase (3), the tectonic influx is smaller than the erosional outflux, leading to crustal thinning, topographic decay and a decrease in denudation rates with time. Material-particle paths (dotted lines with arrows) and isotherms (continuous lines) are schematically indicated for the steady-state case. The graph at the base shows a highly idealised 'lag-time plot' of the detrital thermochronological age versus the stratigraphic age for samples collected from the foreland of a mountain belt experiencing these evolutionary stages. Diagonal lines represent contours of equal lag time. The shaded arrow with dots indicates the predicted evolution of detrital thermochronological age with time: lag times decrease during the constructional phase (1), remain constant during steady state (2) and increase during the destructional phase (3). Compare this with the lag-time plot of real data shown in Figure 9.4 later.

9.1 The basic approach

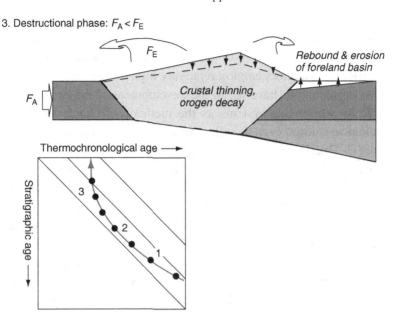

Fig. 9.2. (cont.)

ceases to grow; the mechanics of the system favours exhumation (at a constant rate) rather than crustal thickening or outward growth. Eventually, change in tectonic forcing conditions leads to a decrease of the tectonic mass influx into the system as convergence slows down. As a response, denudation rates will also decrease as topography decays asymptotically during the 'destructional' phase.

The application of detrital thermochronology thus significantly extends the temporal range over which the evolution of a mountain belt can be studied. The spatial resolution of the method, in contrast, is limited. When the method is applied to modern river sediments, a detrital sample integrates the recent exhumation history of the entire drainage basin (or at least of those parts where rocks that contain the mineral analysed crop out) upstream of the sampling site; as shown by Bernet et al. (2001, 2004), the detrital grain ages faithfully record the extant thermochronological age distribution within the upstream catchment. In 'fossil' sediment samples, however, the contributing source area may be ambiguous. Optimum application of detrital thermochronology consequently requires the analysis to be combined with provenance studies such as heavy-mineral petrography (e.g., Ruiz et al., 2004), geochemical data (e.g., Spiegel et al., 2004), or U–Pb geochronology (e.g., Carter and Bristow, 2000; Rahl et al., 2003), in order to constrain the sedimentary source areas.

Because of the potentially large dispersion in thermochronological ages contained in a single detrital sample (see Section 9.2) the method requires relatively large numbers of individual grains to be dated with sufficient precision to distinguish different age groups (Vermeesch, 2004). Thermochronological analysis of detrital samples is therefore restricted to methods that allow high-quality single-grain ages to be measured routinely. In order to avoid potential interpretational problems due to partial resetting of samples when they were buried in the sedimentary basin, most authors employ relatively high-closure-temperature systems. The above considerations have led to the development of the zircon fission-track and ^{40}Ar/^{39}Ar white-mica systems as the methods of choice for detrital thermochronological studies, even though other methods (notably apatite fission-track and zircon (U–Th)/He measurements) are also being developed. The application of the 'external-detector' method (cf. Section 3.3) allows single-grain fission-track dating; single-grain ^{40}Ar/^{39}Ar fusion dating of micas is possible in most noble-gas mass spectrometers. As for *in situ* studies, the characteristics of the study area and the study objectives will determine the method of choice: high-temperature systems (^{40}Ar/^{39}Ar, zircon fission track) will have long 'memories' and not suffer resetting in the basin, but at the cost of relatively long integration times and therefore low temporal resolution; in low-temperature systems (apatite fission track, (U–Th)/He) these relative advantages and drawbacks are reversed.

9.2 Deconvolution of detrital age distributions

Single grains within a given detrital sample will generally exhibit considerably more spread in thermochronological ages than will *in situ* bedrock samples, for a number of reasons. Firstly, the source areas may experience a range of different exhumation rates and therefore supply grains with varying thermochronological ages. Secondly, the detrital grains come from variable sources and may be characterised by variable kinetic behaviour, so that their effective closure temperatures are also variable. For the same reason, partial resetting due to burial in the sedimentary basin after deposition may affect grains with different kinetics variably. Finally, even if it were possible to obtain an unreset sample with uniform annealing or diffusion characteristics from a uniformly eroding catchment, this would still be expected to exhibit a spread in ages because of basin hypsometry and the fact that thermochronological ages generally increase with elevation (see Section 6.2) (Stock and Montgomery, 1996; Brewer *et al.*, 2003).

Consequently, a detrital grain-age distribution will generally be a *mixed* distribution and contain more than one *component* distribution, that is, different groups of grains that are each characterised by a common thermal history and kinetic response (Figure 9.3). The mean age of a mixed distribution does not provide any useful measure; the challenge is therefore to deconvolve the detrital grain-age distribution into its component distributions. Several questions need to be answered: how many components are represented in the mixed distribution; what are their mean component ages (and associated standard errors); and what are

9.2 Deconvolution of detrital age distributions

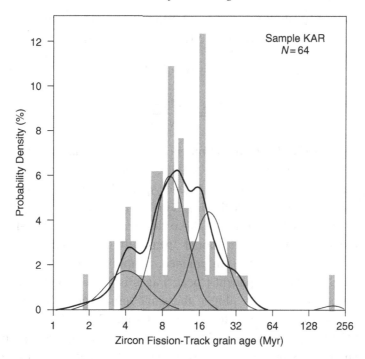

Fig. 9.3. Detrital zircon fission-track ages from a present-day sediment sample from the Karnali River; this is the main river of western Nepal and has a very large (nearly 45 000 km^2) drainage area encompassing most of the western Nepal Himalaya. The figure shows the 64 single-grain ages from this sample as a histogram (grey bars) and a probability-density plot (thick continuous line). The grain ages were deconvolved into four component populations (thin continuous lines) using a binomial peak-fitting approach (*Binomfit*; Stewart and Brandon (2004)); the peak ages, standard errors and relative component sizes are 4.1 ± 1.1 Myr (18%), 9.3 ± 1.3 Myr (46%), 19.1 ± 2.9 Myr (35%) and 216.4 ± 150 Myr (2%), respectively.

their relative weights or contributions to the total distribution? Several approaches to solving these questions have been suggested, mostly developed in application to detrital fission-track analyses (Hurford *et al.*, 1984; Galbraith and Green, 1990; Galbraith and Laslett, 1993; Brandon and Vance, 1992; , Brandon, 1996) although they apply equally well conceptually to other thermochronological systems.

As applied to fission-track dating, where N_s and N_i are the spontaneous and induced fission-track counts for each grain in the sample (see Section 3.3), it can be shown (e.g., Galbraith and Laslett, 1993) that, for any component distribution, N_s should be binomially distributed with a distribution parameter, θ, and index, m, given by

$$\theta = \rho_s/(\rho_s + \rho_i)$$
$$m = N_s + N_i$$

(9.1)

where ρ_s and ρ_i are the spontaneous and induced track densities, respectively. The estimator, $\hat{\theta}$, and standard error, $\mathrm{SE}(\hat{\theta})$, for this distribution are given by

$$\hat{\theta} = N_s/(N_s + N_i)$$

$$\mathrm{SE}(\hat{\theta}) = \sqrt{\frac{\hat{\theta}(1-\hat{\theta})}{N_s + N_i}} \quad (9.2)$$

and can be transformed into the component 'peak' age and its standard error, respectively (Galbraith and Laslett, 1993; , Brandon, 1996). If all grains are characterised by large track counts, N_s and N_i, the binomial distribution can be closely approximated by a normal (Gaussian) distribution (Brandon, 1996).

Therefore, a population of ages can be deconvolved into a best-fit group of component distributions by a binomial 'peak-fitting' approach (Galbraith and Laslett, 1993; , Brandon, 1996). In practice, an initial estimation of the number of components and the component ('peak') ages is made (for instance from the peaks in a probability-density plot of the single-grain ages; (Brandon, 1996)); using a maximum-likelihood method (Galbraith and Green, 1990; Galbraith and Laslett, 1993) the ages and sizes of the peaks are varied until a best-fit solution is found. The process is repeated for successively larger numbers of peaks until no further statistical improvement is found with the inclusion of additional peaks. The question then becomes that of at what stage adding another peak ceases to be warranted, since an absolute best-fit (but trivial) solution contains as many component ages as there are grains in the population. Stewart and Brandon (2004) propose the use of an F-test to answer this question: the quality of each trial solution is scored using a χ^2 statistic and for each number of component distributions the best-fit solution (lowest χ^2 value) is retained. For two solutions with n and $n+1$ peaks and χ^2 values of the best-fit solutions χ_n^2 and χ_{n+1}^2, respectively, the F-statistic is given by

$$F = (n+1)(\chi_n^2 - \chi_{n+1}^2)/\chi_n^2 \quad (9.3)$$

From the distribution of F, a probability $P(F)$ that the improvement in fit can be produced by random variations alone can be calculated. One can then consider the addition of a new peak to provide a significant improvement in the fit until $P(F)$ reaches some cutoff value (i.e., $P(F) \geq 5\%$). The program *Binomfit*, developed by Mark Brandon and described by Stewart and Brandon (2004) offers a user-friendly environment in which to deconvolve detrital fission-track ages using this approach, and can be downloaded from http://www.geology.yale.edu/~brandon/Software/FT_PROGRAMS/BinomFit/index.html. It was used to deconvolve the single-grain age population shown in Figure 9.3.

Deconvolving a detrital grain-age distribution into its component populations does not in itself, however, provide a means of interpreting the significance of the component peak ages. A crucial parameter is the relative magnitude of the component peak ages with respect to the depositional age. If all age peaks are at least as old as the depositional age of the sample, the data can be interpreted as coming from source areas with varying denudation rates, the youngest peak age representing the most rapidly denuding source area. If the younger component ages are more recent than the depositional age, the sample must have suffered post-depositional partial resetting in the sedimentary basin; the youngest age peak may represent a thermal event within the basin. Note that it is the component age peaks rather than the single-grain ages that are of interest here; because of the relatively large errors in the single-grain ages, a few grains dated at younger than the depositional age may exist even in non-reset samples. A final question concerns the relative magnitudes of the peaks: if the sediment is well mixed, these may represent the relative contributions of source areas with variable denudation rates. However, if the short-term spatial distribution of denudation rates is decoupled from the long-term distribution (for instance because landsliding or other highly stochastic erosion processes are important in the catchment), the relative peak sizes might not provide relevant information on the distribution of long-term denudation rates (e.g., Bernet et al., 2004).

As an example of the lag-time approach, Figure 9.4 shows detrital zircon fission-track data from foreland basin sediments surrounding the European Alps. The plot shows the youngest detrital age peak as a function of stratigraphic age; lag-time contours are also indicated. The data come from Eocene–Miocene 'molasse' sediments in France, Germany and Switzerland (Bernet et al., 2005; Spiegel et al., 2004). Bernet et al. (2005) interpreted their data as indicating a constructional phase (decreasing lag times) from 36 to ~ 27 Myr ago (Eocene–Oligocene), followed by exhumational steady state. Spiegel et al. (2004), in contrast, argue that their data are inconsistent with an exhumational steady state before ~ 14 Myr ago. The conflicting interpretations of the two groups of authors may result from the incorporation of data points that show up as outliers when the data are considered collectively. Conspicuously young minimum peak ages at ~ 30 Myr in the Bernet et al. (2005) data may be related to substantial volcanic input at this time. Old outliers are present in the Spiegel et al. (2004) data at 20 and 13 Myr. These data were collected from proximal deposits at various locations close to the deformation front; one can expect them to show more variability since they integrate much smaller source areas. Taken together, the data show a relatively consistent youngest age component with a lag time of 8–10 Myr from ~ 20 Myr ago onward; older stratigraphic horizons are characterised by more scattered and generally longer lag times for the youngest age peak. One could thus argue that these data

140   *Detrital thermochronology*

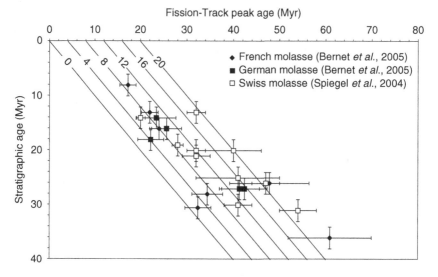

Fig. 9.4. A plot of youngest zircon fission-track component (peak) age as a function of stratigraphic age for samples taken from foreland ('molasse') sediments of the European Alps. Data are from two recent studies, one in the French and German molasse (Bernet *et al.*, 2005), the other in proximal fan deposits in the Swiss molasse (Spiegel *et al.*, 2004). Diagonal lines represent contours of equal lag time as annotated.

record the construction of the European Alps during Eocene–Oligocene times followed by a large-scale denudational steady state, even though the youngest age peak might not consistently be derived from the same area.

9.3 Estimating denudation rates from detrital ages

As discussed above, the temporal variation in lag time can be interpreted qualitatively in terms of increasing, decreasing or steady-state exhumation rates through time. Quantifying denudation rates in the source areas from detrital data more rigorously is challenging, because of the generally large variation in single-grain ages and the multiple factors controlling this variation.

Brandon *et al.* (1998) and Garver *et al.* (1999) proposed a simple one-dimensional analysis to convert lag times of detrital samples into denudation rates of the source area. They approximated the dynamic perturbation of the initial thermal profile of the source area using a steady-state solution for a one-dimensional layer of thickness L (cf. Section 5.1). The upper and lower boundaries of the layer are held at constant temperatures T_s and $T_s + G_0/L$, respectively, where T_s is the temperature at the surface and G_0 is the initial geothermal gradient, which is assumed constant (i.e. constant conductivity, no heat production). The denudation rate is held constant at $\dot{E}$. The steady-state temperature profile for this case is given by Equation 5.8 (given here in its dimensional form):

$$T(z, \dot{E}) = T_s + G_0 L \frac{1 - e^{-\dot{E}z/\kappa}}{1 - e^{-\dot{E}L/\kappa}} \quad (9.4)$$

where z is depth and κ is the thermal diffusivity of the crust. As discussed in Chapter 2, the effective closure temperature of a sample T_c can be calculated as a function of its cooling rate (Dodson, 1973). An approximate relationship for this effect is given by Equation (2.21):

$$T_c = \frac{E_a}{R \ln\left(B \frac{RT_c}{E_a \dot{T}}\right)} \quad (9.5)$$

where R is the gas constant, E_a is the activation energy of the system and B is a proportionality constant that combines the parameters (AD_0/a^2) in Equation (2.21). The cooling rate at closure is a function of the exhumation rate of the sample and the vertical gradient in the temperature profile at T_c:

$$\dot{T}|_{(T_c)} = \frac{\partial T}{\partial z} \frac{\partial z}{\partial t} = \frac{\dot{E}^2}{\kappa}\left(\frac{G_0 L}{1 - e^{-\dot{E}L/\kappa}} - (T_c - T_s)\right) \quad (9.6)$$

Equation (9.4) is inverted to define z_c, the depth at which closure occurs:

$$z_c = -\frac{\kappa}{\dot{E}}\left(1 - \frac{T_c - T_s}{G_0 L}(1 - e^{-\dot{E}L/\kappa})\right) \quad (9.7)$$

A final equation relates z_c to the thermochronological age τ and the constant exhumation rate $\dot{E}$:

$$z_c = \dot{E}\tau \quad (9.8)$$

Combining Equations (9.5)–(9.7) with Equation (9.8) gives two equations that can be solved numerically for the two unknowns, $\dot{E}$ and T_c (Brandon et al., 1998). The enclosed Fortran code 'lagtimetoexhum.f' implements this iterative solution. As input, it requires the diffusion parameters E_a and D_0/a^2, as well as the thermal parameters T_s, G_0, L and κ. Its outputs include the closure temperature T_c, closure depth z_c and denudation rate $\dot{E}$ as a function of thermochronological age (Figure 9.5).

As the exhumation rate experienced increases, samples move from closure to exposure more rapidly, and lag times decrease accordingly. This trend is enhanced by the increased perturbation of regional thermal structure associated with higher exhumation rates (as discussed in Section 5.1), which physically moves the relevant closure isotherm closer to the surface, producing the exponential form of the exhumation-rate–age relationship shown in Figure 9.5.

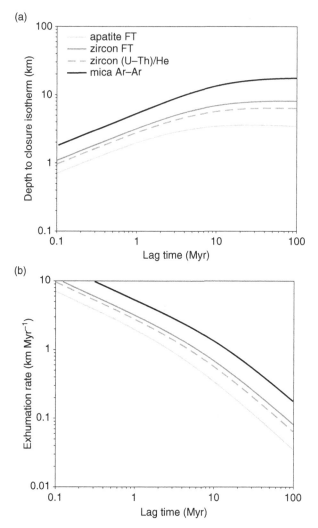

Fig. 9.5. The depth to the closure isotherm z_c and exhumation rate $\dot{E}$ as functions of the lag time for four different thermochronometers. Diffusion parameters for zircon (U–Th)/He and muscovite $^{40}Ar/^{39}Ar$ are taken from Tables 3.4 and 3.2, respectively, for grain sizes of 100 μm; parameters for zircon and apatite fission-track (FT) ages were fitted to the available experimental annealing data by Brandon et al. (1998): $E_a = 186\,kJ\,mol^{-1}$, $B = 9.8 \times 10^{11}\,s^{-1}$ for apatite FT annealing; $E_a = 208\,kJ\,mol^{-1}$, $B = 10^8\,s^{-1}$ for zircon FT annealing. The thermal parameters used were $T_s = 10\,°C$, $G_0 = 25\,°C\,km^{-1}$, $L = 30\,km$ and $\kappa = 25\,km^2\,Myr^{-1}$.

Garver et al. (1999) used this approach to calculate the mean denudation rate of the source area from the single-grain age distribution of a detrital sample (Figure 9.6). They converted each single-grain age into a corresponding denudation rate to construct a probability-density plot of source-area denudation rates.

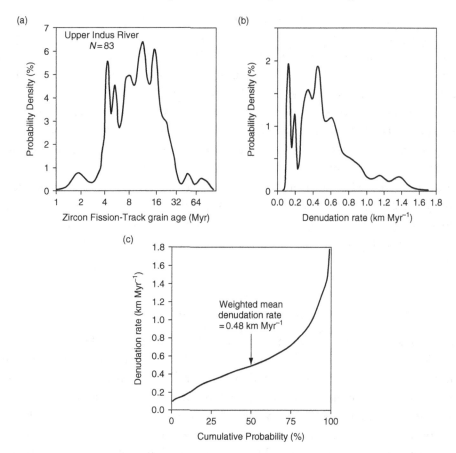

Fig. 9.6. Transformation of a detrital single-grain thermochronological age distribution into a corresponding distribution of source-area denudation rates. (a) Detrital zircon fission-track data from modern Indus River sediments, Pakistan, plotted as a probability-density curve. (b) The weighted probability-density curve for calculated denudation rates; each single-grain age was transformed into a denudation rate using Equations (9.4)–(9.8) and the resulting probability-density function was weighted by dividing over denudation rate and renormalising. (c) The same result plotted as a weighted cumulative probability function. Modified from Garver et al. (1999); original data from Cerveny et al. (1988). Reproduced with permission by the Geological Society of London.

In order to correct for the higher sediment yield of rapidly denuding areas (which would be proportionately over-represented in the detrital sample) the probability represented by each grain was renormalised by dividing it by the corresponding denudation rate.

Note that this approach relies on rather significant simplifications, which may strongly affect the predicted denudation rates. Firstly, since the calculation uses the simplified closure-temperature concept, all predicted denudation rates are necessarily temporal averages over the time span from closure to deposition (i.e. the lag time); any variations in denudation rate over this time span cannot be resolved. As an example, the few grains with 'pre-Himalayan' (i.e., >50 Myr) zircon fission-track ages in Figure 9.6(a) translate into a significant peak of low denudation rates (<100 m Myr^{-1}) in Figure 9.6(b). However, the interpretation that these samples record slow and continuous exhumation at these rates since the onset of Himalayan collision is geologically unfeasible; rather, they reflect a long and possibly complex history of shallow burial, more recent exhumation and possible sedimentary recycling. Secondly, apart from the temporal averaging, all variations in detrital thermochronological ages are interpreted as spatial variations in source-area denudation rates; the calculation of denudation rates is strictly one-dimensional. Neither the effect of topography nor that of the integration of lateral variation in regional character on thermochronological ages (Chapters 6 and 10) are taken into account.

9.4 Estimating relief from detrital ages

At the other conceptual extreme, instead of explaining the detrital grain-age variation as a consequence of variable denudation rates in the source area, an alternative approach lies in considering that denudation rates in the source area are constant and that the grain-age variation is a result of source-area relief. In that case, the (paleo)relief of the source area can be estimated from the probability-density function of the detrital grain-age distribution, by deconvolving it with a known or predicted thermochronological age–elevation relationship in the source area (Stock and Montgomery, 1996; Brewer *et al.*, 2003).

The approach is outlined in Figure 9.7. In order to constrain paleorelief, the age–elevation gradient can be modelled by using the methods derived by Stüwe *et al.* (1994) and Mancktelow and Grasemann (1997) and outlined in Chapter 6. Multiplying the probability-density function of detrital single-grain ages by the age–elevation gradient then yields the source-area relief. In principle, the combination of a detrital component age peak and the peak width can then be used to constrain both the exhumation rate and the relief of the source area.

Figure 9.8 shows predicted age ranges for a high-temperature thermochronometer (^{40}Ar – ^{39}Ar on white mica) and a low-temperature (apatite fission-track) thermochronometer, for a range of exhumation rates and relief values. The thermochronological ages are predicted in a manner similar to that in the previous section but taking a two-dimensional steady-state thermal structure into account: the thermal structure for a given exhumation rate ($\dot{E}$), topographic wavelength (λ) and relief amplitude (z_0) is calculated using the approach of Mancktelow and Grasemann (1997) (cf. Equation (6.11)). From this, the temperature and cooling

9.4 Estimating relief from detrital ages

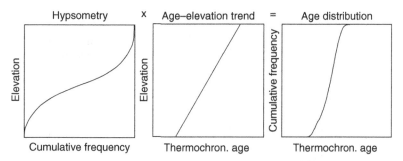

Fig. 9.7. The concept of estimating (paleo)relief from detrital thermochronological data: convoluting the source-area hypsometry with the age–elevation trend should result in the detrital single-grain age distribution, for spatially constant exhumation rates and steady-state topography. Modified from Brewer *et al.* (2003).

rate beneath the summits and valley bottoms are predicted as functions of depth and the closure temperature is estimated from Equation (9.5). Combining the two provides an estimate for the cooling depth (z_c) beneath summits and valleys; dividing this depth by the exhumation rate provides the predicted summit and valley ages. Figure 9.8 shows that, for exhumation rates $\leq 1.5 \, \mathrm{km \, Myr^{-1}}$ and relief $> 2 \, \mathrm{km}$, the relief of the source area should result in resolvable variations in detrital ages, both for low- and for high-temperature systems. Because of the stronger perturbation of isotherms near the surface, absolute differences between summit and valley ages are smaller for low-temperature than for high-temperature systems, but the relative variations are larger.

As for the previous approach, however, the predictions of source-area relief from detrital age data are limited by strong simplifying assumptions. Notably, in order for the analysis to make sense, exhumation rates should be constant throughout the source area and relief should be in steady state. Moreover, the approach assumes that single-grain ages can be obtained with sufficient precision to constrain source-area relief meaningfully and that kinetic variation between grains is negligible. Finally, this approach also requires that the sediment sample is well mixed so that the probability density distribution faithfully represents the age distribution in the source area.

The limitations of both end-member approaches are illustrated in Figure 9.9, which shows detrital apatite fission-track data from a present-day river sand in the western Alps, where both the hypsometry and the age–elevation gradient of the source area are known, the latter from an independent age–elevation profile. Interpreting the detrital age data in terms of variations in source-area exhumation rates leads to an underestimation of the mean denudation rate (relative to the rate obtained from the age–elevation relationship) whereas interpreting the data

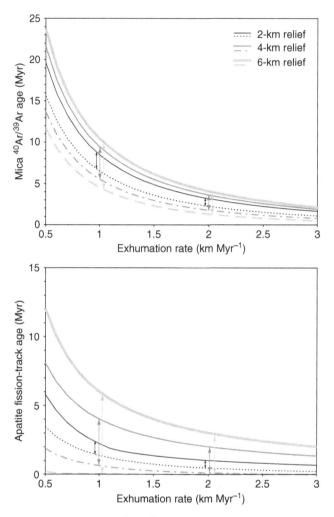

Fig. 9.8. Predicted white-mica $^{40}Ar/^{39}Ar$ and apatite fission-track ages at summits (continuous lines) and valleys (dashed/dotted lines) for a range of denudation rates and three different relief amplitudes ($z_0 = 2$, 4 and 6 km). Ages are calculated by combining the calculations of Figure 9.5 with the two-dimensional steady-state thermal structure under periodic topography (Equation (6.11)). The wavelength of topography $\lambda = 10$ km; other parameters are as in Figure 9.5. Double arrows indicate predicted age ranges for denudation rates of 1 and 2 km Myr^{-1}, respectively. Note the difference in age scale between plots.

in terms of a steady-state relief exhuming at a constant rate does not provide a reasonable fit to the source-area hypsometry. This result implies that (1) the sediment is not well mixed, (2) some of the single-grain age variation is due to kinetic variation between the grains, (3) denudation rates are not uniform or (4) the relief is not in steady state. A qualitative interpretation of the data

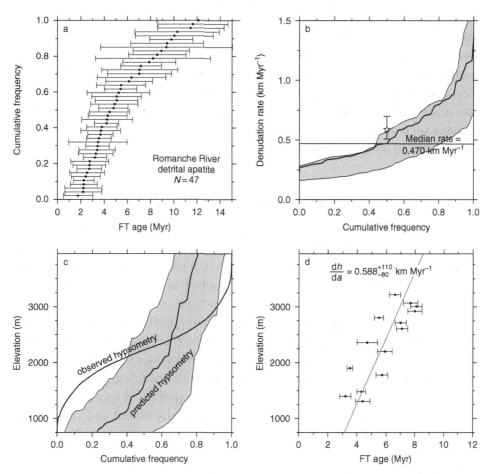

Fig. 9.9. An example of using detrital thermochronological data to predict source-area denudation rates or relief. (a) The cumulative distribution of 47 single-grain apatite fission-track (FT) ages (and their associated 1σ errors) from present-day sediment from the Romanche River, which drains the northern part of the Pelvoux–Ecrins massif in the French western Alps. (b) The weighted cumulative probability density function of source-area exhumation rates calculated from these data, using the approach outlined in Section 9.3; shading indicates the 1σ error in the predicted rates. The median source-area exhumation rate is $470\,\text{m}\,\text{Myr}^{-1}$; the star indicates the denudation rate inferred from the age–elevation profile (shown in (d)) of $590^{+110}_{-80}\,\text{m}\,\text{Myr}^{-1}$. (c) Predicted source-area hypsometry, with 1σ error, calculated from the single-grain ages in (a) and the age–elevation relationship in (d), compared with the observed hypsometry of basement rocks containing apatite in the drainage basin upstream of the sample site. (d) Apatite FT ages from an age–elevation profile on La Meije peak in the upper Romanche drainage basin. The best-fit age–elevation profile ($h = 0.0017a - 1.8058$; $r^2 = 0.54$) indicates a denudation rate of $588^{+110}_{-80}\,\text{m}\,\text{Myr}^{-1}$ (95% confidence limits).

predicts that source-area relief is increasing by focussed incision of the valleys, leading to the observed predominance of 'young' ages (≤ 4 Myr) in the detrital sample. These young ages would then reflect variable denudation rates, whereas the spread in older ages could be explained by invoking the observed age–elevation gradient. Optimum analysis of detrital grain-age populations, taking into account the possibility of non-uniform denudation rates and transient relief, requires sophisticated modelling approaches such as those outlined in Chapter 7, combined with inversion techniques as discussed in Chapter 8.

9.5 Interpreting partially reset detrital samples

Whereas the high-temperature zircon fission-track or mica Ar–Ar systems are generally better suited for monitoring long-term variations in exhumation rate of the source area, the lower-temperature apatite fission-track and (U–Th)/He systems allow testing for shorter-term variations in exhumation rates. Because of their relatively low closure temperatures, however, these systems are also correspondingly more sensitive to the thermal evolution of the basin itself. Typically, stratigraphically higher samples that have not been buried deep enough for partial resetting to occur will retain a source signal, whereas deeper samples will record the burial and exhumation history of the basin sediments themselves (e.g., Rohrman *et al.*, 1996). Therefore, as samples from deeper downsection are analysed, their information content on source-area denudation is gradually erased and replaced by information on the burial and exhumation history of the sedimentary basin.

The problem of retrieving information from partially reset samples has been analysed quantitatively for the apatite fission-track system, which is widely used by the petroleum industry for studies of the thermal evolution of sedimentary basins (Rohrman *et al.*, 1996; Carter and Gallagher, 2004). Given minimal controls on the depositional history and kinetic properties of the apatites studied, both the pre-depositional (source area) and the post-depositional (sedimentary basin) thermal history can, in principle, be recovered from detrital apatite fission-track samples by inverting their track-length distributions (Carter and Gallagher, 2004). In practice, however, the potentially large variations in source-area denudation rates, as well as in the annealing kinetics of the sampled grains, make this approach challenging. Nevertheless, important information on the thermal history both of the source area and of the basin can be gained by inspecting the component ('peak') ages if data are sampled continuously downsection.

An example of such an approach is shown in Figure 9.10, where detrital apatite fission-track (AFT) data from the Karnali River section (western Nepal) have been plotted as a function of the initial stratigraphic depth of the samples. The Karnali section exposes Miocene–Pliocene (16–5-Myr-old) Himalayan foreland

9.5 Interpreting partially reset detrital samples 149

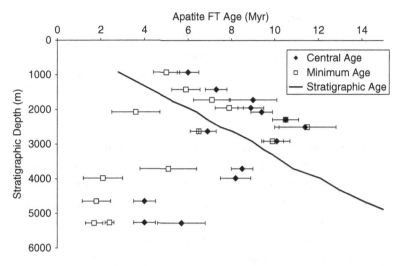

Fig. 9.10. Variations of apatite fission-track central age (black diamonds) and minimum peak age (white squares) with stratigraphic depth in the Karnali River section through Siwalik sediments in western Nepal. The continuous black line is the stratigraphic age (after Gautam and Fujiwara, 2000). The initial stratigraphic depth was estimated by extrapolating the average sedimentation rates to the present day. Note the constant ~2 Myr lag time between stratigraphic and minimum apatite fission-track (FT) ages in the upper part of the section (≤2500 m) and the consistent 2 Myr minimum age for the lowest four samples, providing an estimate for the final exhumation of the section along the Main Frontal Thrust.

basin sediments of the Siwalik group (e.g., Gautam and Fujiwara, 2000). These are exposed at present because of tilting and exhumation along the Main Frontal Thrust, the currently active frontal thrust of the Himalayan system. Samples from the upper part of the section (down to a paleo-depth of ~2500 m) have AFT ages that increase downsection; both the central age and the peak age of the youngest component (the 'minimum' age) are older than the stratigraphic age. These samples have not been annealed since deposition and retain the source-area information. Their minimum AFT age displays a constant lag time of ~2 Myr since ~7.5 Myr ago, which is indicative of maximum source-region exhumation rates of ~1.5 km Myr^{-1}, similar to the longer-term rates measured by detrital zircon fission-track analysis of these samples.

For the stratigraphically lower samples, AFT ages decrease downsection and are younger than the stratigraphic age: these samples have been partially annealed during burial in the basin. The transition from unannealed to partially annealed samples represents the exposed top of the 'fossil' apatite partial annealing zone (PAZ) at ~60 °C (compare with Figures 3.8 and 3.10). The paleo-geothermal gradient can be inferred from the estimated structural depth of this transition; in the case of Figure 9.10, a pre-exhumational geothermal gradient of 15–20 °C km^{-1} within the basin is inferred, which is notably in accord with vitrinite reflectance data and present-day heat-flow measurements in wells from the Ganges foreland basin. Alternatively, if the geothermal gradient within the basin is known, the depth to the top of the paleo-PAZ can be used to infer the amount of denudation following the approach outlined in Figure 1.6 (e.g., Cederbom et al., 2004).

The base of the paleo-PAZ is represented by the transition from partially to fully annealed samples; for cases in which final exhumation of the samples was rapid, fully annealed samples will be characterised by constant thermochronological ages downsection that represent the age of final exhumation (cf. Figures 3.10 and 1.6). The base of the paleo-PAZ is not exposed in the Karnali section shown in Figure 9.10 but the minimum ages can be used to place limits on the timing of final exhumation. In particular, the annealed AFT samples from the lowermost part of the section have a consistent minimum age peak of 2 ± 0.4 Myr and therefore suggest that the onset of exhumation of this part of the Siwaliks along the Main Frontal Thrust occurred at that time. Using a similar approach on well samples, Cederbom et al. (2004) showed that the molasse foreland basin of the central Alps records 1–3 km of exhumation since 5 Myr ago, which they relate to isostatic rebound of the basin due to widespread and rapid denudation of the mountain belt.

10
Lateral advection of material

> *When one is applying a physical interpretation to thermochronological data as outlined here, the geographical distribution of deformation, denudation and thermal structure, and the interaction of particles with these laterally variable parameters justify careful consideration. In this chapter, we will investigate the consequences of lateral variation in denudation rates and lateral motion of material points on thermochronological age distributions.*

10.1 Lateral variability in tectonically active regions

Thermochronological ages at the surfaces of eroded regions reflect an integration of the denudation rates experienced between the isotopic closure of a sample and its subsequent exposure. Crustal structure and kinematics (Beaumont *et al.*, 1992, 1996; Koons, 1994), thermal structure (Koons, 1987; Lewis *et al.*, 1988; Hyndman and Wang, 1993; Batt and Braun, 1997; Ehlers *et al.*, 2001) and regional exhumation rates (Wellman, 1979; Zeitler *et al.*, 1982; Brandon *et al.*, 1998; Ehlers *et al.*, 2003) can all have strong lateral gradients across tectonically active regions. Any material traversing this variable framework is thus subject to an accompanying range of conditions during its residence within the deforming region that will influence the development of the apparent thermochronological age.

Material denudation, and the uplift and deformation with which it is commonly associated, are driven, fundamentally, by lateral motion between crustal blocks. Far from being an exceptional occurrence, lateral motion of material dominates the overall kinematics experienced in many tectonically active regions, with samples experiencing tens or even hundreds of kilometres of lateral translation during progressive uplift and exhumation from mid-crustal depths (Beck, 1991; Jamieson *et al.*, 1996; Walcott, 1998; Brandon and Vance, 1992; Brandon *et al.*, 1998), and

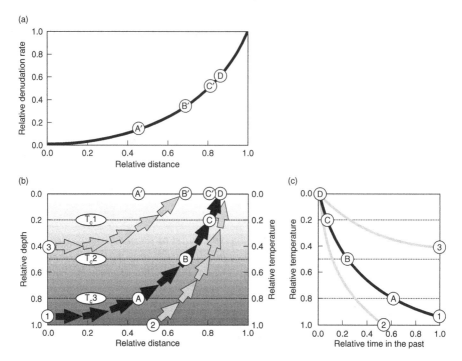

Fig. 10.1. The thermochronological significance of material convergence relative to the denudational and thermal framework of a deforming region. Note that, although figurative, the processes and effects illustrated are derived from published (Batt and Braun, 1997, 1999; Batt *et al.*, 2001) and unpublished numerical modelling work of the authors. (a) The assumed variation in erosion rate across the area of interest. (b) The regional thermal structure, and the material path taken by three selected particles during their passage through the deforming region. Arbitrary closure isotherms are shown for three nominal thermochronometers, as discussed in the text. (c) Variations in temperature experienced over time by the three particles illustrated in panel (b). Note the progressive acceleration in cooling rate. This is caused by spatial, rather than temporal, variation in exhumation rate (panel (a)), and thus illustrates the caution required in attaching dynamic significance to such features.

hence integrating tectonic and surface conditions over potentially great distances (Figure. 10.1).

10.2 Exhumation and denudation in multi-dimensional space

As defined in Chapter 1, exhumation refers to the unroofing history of an actual rock, defined as the vertical distance traversed relative to the Earth's surface, whereas denudation, taken in the broad generic sense of Ring *et al.* (1999), relates to the removal of material at a particular point at or under the Earth's surface, by tectonic processes and/or erosion, and is more correctly viewed as a measure of material flux.

In the one-dimensional case in which the regional character is uniform, or particles are exhumed purely vertically, the terms denudation and exhumation are practically synonymous, except for the subtle distinction in the frame of reference. For cases in which lateral transport plays an important role, in contrast, the exhumation record derived from thermochronological constraints is not directly interchangeable with spatial parameters such as erosion or tectonic unroofing (Morris et al., 1998) (see Figure 10.1).

10.3 Consequences of lateral motion for thermochronology

Integrated effects on individual ages

The potential thermochronological consequences of lateral motion relative to a variable tectonic framework are schematically illustrated in Figure 10.1. Material is assumed to be traversing a region experiencing deformation and erosion where denudation rates vary laterally. For simplicity, the thermal structure is shown as uniform, with a linear geothermal gradient, but the detail of this is not critical to the basic phenomena under discussion here. Sample 1 (the main path illustrated) cools progressively through the closure temperatures of three arbitrary thermochronometers during its exhumation, experiencing effective closure at points A, B, and C respectively, with A', B' and C' marking the corresponding points at the surface directly above each of these sites. Upon its eventual exposure at the surface, the apparent ages of sample 1 for each of these three chronometers reflect the average exhumation rate experienced between the relevant point of closure (A, B, or C) and exposure of the sample at point D. Owing to the spatial variation in denudation rate, the exhumation rate experienced varies through time as the sample traverses the deforming region. As a consequence, none of the chronometers indicated will directly reflect the denudation rate either at point D or at the location at which its respective closure occurred.

Consequences for spatially dispersed datasets

Lateral variation in the character of exhumation paths has equally significant consequences for the integrated interpretation of spatially dispersed thermochronological data, where physical significance is ascribed to variations in ages between data collected at different localities across a region.

When one is sampling age–elevation profiles (cf. Sections 1.2 and 6.2), tectonic behaviour is interpreted from age variations along transects in areas of high relief (Wagner and Reimer, 1972; Wagner et al., 1977; Fitzgerald and Gleadow, 1988; Fitzgerald et al., 1995). Conceptually, this approach is based on treating samples as if they all came from a single column of rock, but, in practice, the collection of samples dispersed over 1000 m or more of elevation difference usually involves lateral separation of material by kilometres to tens of kilometres, even in areas of intense relief.

At longer spatial wavelengths, authors of some tectonic studies ascribe physical or temporal significance to geographic variations in the apparent age of samples collected across deformed and exhumed regions at the scale of entire orogens (Brandon and Vance, 1992; Tippett and Kamp, 1993; Batt et al., 2000, 2004; Cockburn et al., 2000; Ehlers et al., 2003).

These approaches benefit from considering lateral variation in kinematics since each point exposed on the surface experiences a different exhumation path, and hence reflects a separate set of interactions with the variable thermal and denudational character of the region. This phenomenon is illustrated in Figure 10.1 by the contrasting histories experienced by samples 1, 2 and 3. Although the regional framework of denudation rates remains stable, these samples experience differing exhumation histories as they traverse the region between isotopic closure and eventual exposure.

This effect introduces ambiguity into the interpretation of spatial variability in thermochronological ages, since any variation observed between geographically separate samples can reflect either temporal variations in behaviour (e.g., diachronous deformation episodes or changes in exhumation rate over time) or variations in the kinematic framework (e.g., structural variation or laterally variable surface processes), or some combination of the two.

The significance of such 'tectonic assembly' of varying thermochronological trends has long been recognised in modelling studies examining regional tectonic evolution (Stüwe et al., 1994; Jamieson et al., 1996, 2002; Batt and Braun, 1997; Harrison et al., 1998), and has led to a wide acceptance of the importance of explicit structural control as an element in thermochronological studies (Stockli et al., 2001; Ehlers et al., 2003).

10.4 Scaling of lateral significance with closure temperature

The higher the temperature at which a system closes, the deeper in the crust and the earlier closure occurs, the greater the lateral distance that will subsequently be travelled prior to exposure at the surface, and the wider the potential range of lateral variation which could thus be incorporated into the observed age (see Figure 10.1). Conversely, the lower the temperature of closure, the less sensitive a thermochronometer is to lateral variation in orogenic character, and the more sensitive it is to local denudation conditions, surface processes and topographic effects. When subject to lateral motion through the denuding region, it follows that different thermochronometers do not just constrain varying timescales or levels of sensitivity, but also provide insight into fundamentally different aspects of the character and evolution of tectonically deformed regions.

10.5 Evaluation of the significance of lateral variation

The consequences of lateral variations in particle paths for the development of the thermochronological record do not mean that one-dimensional interpretations cannot provide valid tectonic insights in some, or indeed many, cases – but the potential impacts of spatially variable structural and erosional character first need to be considered in order to evaluate their relative significance.

Regional estimation of significance: the η factor

Before embarking on time- and processing-intensive deconvolution of the thermotectonic implications of individual samples, a useful first step is to consider the approximate level of influence that lateral motion might be expected to exert within a particular dataset. A simple guide to the potential consequences of lateral particle motion during the buildup of thermochronometric ages can be derived from comparison of the spatial variability in denudation rate $f_{\Delta\epsilon}$,

$$f_{\Delta\epsilon} = \left(\frac{\epsilon_{max} - \epsilon_{min}}{\epsilon_{max}}\right) \Big/ \lambda_{\Delta\epsilon} \tag{10.1}$$

where ϵ_{max} and ϵ_{min} are the maximum and minimum denudation rates across the region and $\lambda_{\Delta\epsilon}$ is the length scale over which this variation occurs. The horizontal length scale λ_X over which the sample has travelled between closure at depth and exposure at the surface (Figure 10.2) is

$$\lambda_X = v_h t_s \tag{10.2}$$

where v_h is the rate of horizontal motion across the region and t_s is the sample age (Batt and Brandon, 2002). Multiplied together, these give a dimensionless number η,

$$\eta = \lambda_X f_{\Delta\epsilon} \tag{10.3}$$

that indicates the relative variation in exhumation rate potentially experienced by a sample between thermochronological closure and its exposure at the surface. If either the variability in denudation rate ($\epsilon_{max} - \epsilon_{min}$) or the horizontal velocity v_h is small, η tends towards zero, and neglecting the thermochronological consequences of lateral motion may be reasonable. Conversely, high values of η approaching or even exceeding 1, that is, reflecting a nominal 100% variation in exhumation rate between closure and surface exposure, indicate that a wide range of exhumation rates could potentially be integrated by the given thermochronometer, with corresponding influence on the observed ages.

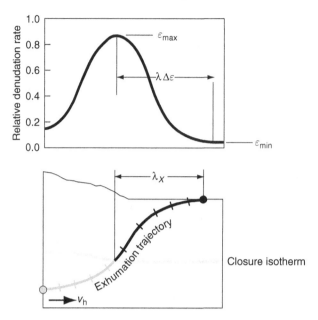

Fig. 10.2. An illustration of the physical parameters relevant to evaluation of the significance of lateral transport in interpreting thermochronological data. A particle passing through a system experiencing exhumation will integrate lateral variation in character over a characteristic length scale, λ_X, that can be derived by combining the average lateral rate of motion through the system with the total time recorded between closure and exposure of the particle given by its apparent age for a given chronometer. The relative significance of this length scale is assayed by comparison with the wavelength of spatial variation in the relevant parameter (in the case illustrated, exhumation rate) within the system, and the total variation in conditions within this range.

Deconvolving lateral effects on the thermochronological record

In the absence of strong dynamic forcing or the transient effects of magmatic intrusions, temperature increases with depth in the crust, with isotherms approximately parallel to topography near the surface and becoming progressively more subdued with depth (as treated in detail in Chapters 4–6). It follows that, despite its consequences for the integrated thermal history experienced by material traversing a tectonically active region, long-term lateral motion is notoriously difficult to isolate in the thermochronological record, since such movement is largely parallel to broad-scale thermal structure and thus does not produce a significant and predictable cooling effect (see, for instance, Figure 13.4 later).

The most effective resolution to this ambiguity is derived through numerical simulations that incorporate the two- and three-dimensional kinematics of tectonic deformation and denudation, and thus allow the thermochronological record to be interpreted in the context of the overall motion experienced.

10.5 Evaluation of the significance of lateral variation

An example of such a simulation will be treated in Chapter 13. In brief, geological and structural controls, together with requirements for conservation of material, enable one to make a reasonable approximation of the past and present mode of deformation of many deformed and denuded regions. Numerical models based on such kinematic frameworks can be used to solve for the evolving thermal structure of the deforming region (Koons, 1987; Beaumont et al., 1996; Batt and Braun, 1997). The physical and thermal coupling of these models enables the tracking of selected material points through the model domain, and the assessment of how a given exhumation path interacts with the evolving thermal structure, enabling thermal histories, and their implications for different thermochronometers, to be calculated for individual particles.

The multi-variate models needed to incorporate the dynamic interplay of heat flow, kinematics and sample mineralogy that goes into producing a thermochronological age cannot usually provide a unique interpretation for specific age data (e.g., Quidelleur et al., 1997). Rather, such models are used to provide insights into how various aspects of tectonic deformation and accompanying denudation are ultimately integrated into the thermochronological record, thus helping to choose between competing hypotheses, and guiding interpretation of the physical causes behind observed regional patterns in thermochronological data (Shi et al., 1996; Beaumont et al., 1996; Batt and Braun, 1999; Batt et al., 2001; Ehlers et al., 2001).

Case study: the Olympic Mountains

The Olympic Mountains (Figure 10.3) are the topographically highest and most deeply exhumed segment of the Cascadia forearc high of western North America. The Olympics were the earliest part of the forearc to become emergent, c. 12 Myr ago (Brandon and Calderwood, 1990), and the mountainous topography of the range has been sustained since that time by continued accretion of material from the subducting Juan de Fuca Plate and within-wedge deformation (Brandon and Calderwood, 1990; Brandon et al., 1998). The Cascadia accretionary wedge (Figure 10.3) is composed predominantly of sedimentary material built up by this progressive accretion (Clowes et al., 1987; Brandon and Calderwood, 1990). This sedimentary wedge underlies most of the offshore continental margin, reaching thicknesses of 30 km, but is sub-aerially exposed only in the Olympic Mountains (Stewart, 1970; Rau, 1973; Tabor and Cady, 1978).

In common with many accretionary complexes, there is a general paucity of age-diagnostic fossils within the sandstones of the Cascadia accretionary wedge. In the absence of reliable paleontological age control, thermochronological data have long been the prime means applied to constraining the tectonic evolution

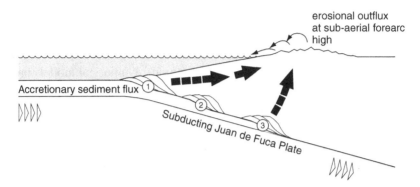

Fig. 10.3. A figurative cross-section illustrating key aspects of the Olympic Mountains segment of the Cascadia accretionary wedge of North America, after Brandon et al. (1998). Deformation of the system is controlled by the influx of material incorporated from the subducting Juan de Fuca slab, and interpretations of the thermochronological record from this region depend critically on the relative significance ascribed to lateral motion through the system during the interval of constraint for a given chronometer. The black arrows illustrate conceptual material-flux paths suggested by competing hypotheses of Olympic orogen kinematics: (1) frontal accretion of the incoming sedimentary section of the Juan de Fuca plate at the toe of the Cascadia wedge (e.g., Davis and Hyndman, 1989); (3) underplating of subducted material at depth beneath the orogen (e.g., Clowes et al., 1987; Brandon and Calderwood, 1990); and (2) some combination thereof.

of the Olympic orogen (Tabor, 1972; Brandon and Vance, 1992; Brandon et al., 1998; Batt et al., 2001; Stewart and Brandon, 2004).

One-dimensional interpretations of thermochronological data from the Olympics (e.g., Brandon et al., 1998) are unable to test the relative merits of the kinematic scenarios proposed for the evolution of the orogen since the differences between competing models lie primarily in the relative importance and orientation of lateral motion of material through the deforming orogen (Batt et al., 2001). In applying coupled dynamic models to the deconvolution and interpretation of fission track and (U–Th)/He data, Batt et al. (2001) were able to highlight the significance of this motion. Figure 10.4 shows a comparison between apatite fission-track ages synthesised by Batt et al. (2001), using a two-dimensional numerical model of the Olympic Mountains, and the one-dimensional analysis used previously to interpret the thermochronological data from the Olympics by Brandon et al. (1998).

Denudation rates across the Olympic Peninsula vary from zero at its western coast to a peak of about $1\,\mathrm{mm\,yr^{-1}}$ in the region of Mt Olympus 30 km to the east (Brandon et al., 1998; Pazzaglia and Brandon, 2001) (Figure 10.3). The models of Batt et al. (2001) predict a horizontal material velocity relative to

10.5 Evaluation of the significance of lateral variation

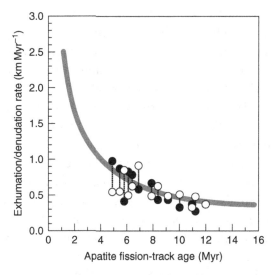

Fig. 10.4. The relationship of modelled fluor-apatite fission-track ages at the surface of the Olympic Mountains to the exhumation rate. The shaded line shows the one-dimensional relationship, calculated after Brandon et al. (1998), for a 20-km-thick crustal layer, assuming a thermal diffusivity of $20\,\text{km}^2\,\text{Myr}^{-1}$ and a thermal structure at equilibrium, as discussed in the text. Paired circular symbols represent the corresponding relationship for individual samples analysed in a two-dimensional model of the Cascadia accretionary-wedge–forearc system (Batt et al., 2001). Open circles mark the denudation rate at the site of a sample's exposure at the surface, with filled circles marking the exhumation rate experienced at the modelled time of closure. After Batt and Brandon (2002). Reproduced with permission by Elsevier.

this denudation framework of about $4\,\text{mm}\,\text{yr}^{-1}$ at the west coast of the Olympic Peninsula, in agreement with the shorter-term estimate of about $3.7\,\text{mm}\,\text{yr}^{-1}$ deduced by Pazzaglia and Brandon (2001) from an offset ~122-kyr-old marine terrace surface. Minimum ages in a large-sample database collected across the orogen are approximately 2 Myr for the apatite (U–Th)/He chronometer, 5 Myr for fission tracks in apatite and 13 Myr for fission tracks in zircon (as summarised in Batt et al. (2001)).

On applying the indicator developed by Batt and Brandon (2002) and discussed in Section 10.5, these parameter values yield η equal to 0.27 for apatite (U–Th)/He ages, 0.67 for apatite fission-track ages and 1.73 for zircon fission-track ages (indicating potentially up to 27%, 67% and 173% variation in exhumation rates between thermochronological closure and exposure at the surface for these three chronometers, respectively). The high value for zircon reflects the fact that predicted lateral translation of samples for this chronometer subsequent to closure exceeds the 30-km length scale of variation in denudation.

In the study of Brandon et al. (1998), the observed apatite fission-track ages were converted into denudation rates by iteratively calculating the closure temperature and the perturbation of the one-dimensional conductive isotherm, using the method outlined in Section 9.3. For the sediment-rich Cascadia accretionary wedge, Brandon et al. (1998) settled on a thermal diffusivity of $\kappa \approx 20\,\text{km}^2\,\text{Myr}^{-1}$ (Brandon and Vance, 1992) and a thickness of $L = 20\,\text{km}$; this value approximates the average thickness of the rear part of the accretionary wedge since middle Miocene times. They used empirically fitted values of the activation energy and diffusivity to model apatite fission-track annealing (cf. Figure 9.5).

Batt et al. (2001) applied an analogous process in two dimensions to predict ages in their coupled thermal and kinematic numerical models of orogenic evolution. They solved for the dynamic effects of material motion and denudation on thermal structure within the deforming Cascadia accretionary wedge, assessed the passage of individual samples through the thermal field, and derived model ages from the resulting thermal history. In one conceptual difference of note between the two approaches, Batt et al. (2001) modelled the actual annealing of fission tracks in samples, following the approach outlined in Section 3.3, rather than a more abstract empirical closure relationship. Given that the closure temperature calculation of Brandon et al. (1998) is derived as a numerical simplification of annealing behaviour, however, the two approaches should yield comparable ages for a given thermal history, except for samples exhumed from within the partial-retention zone for the chronometer in question. Because of the lateral variation in $\dot{\epsilon}$, local denudation rates at the sites where most samples in the two-dimensional model developed by Batt et al. (2001) are eventually exposed differ significantly from the rates to which the local thermal structure was equilibrated at the point of apatite fission-track closure (Figure 10.4). As a result, no simple relationship can be drawn between the age of a sample and the local dynamics of the crust. As shown in Figure 10.4, ignoring this effect and interpreting these apatite fission-track ages under the assumption of one-dimensional behaviour could result in errors in the estimated local denudation rate approaching the theoretical 67% level predicted in the analysis above. This interpretational error would be proportionately magnified for zircon fission-track ages and other chronometers of higher closure temperature (Figure 10.1).

Tutorial 9

The Southern Alps of New Zealand are another orogen for which lateral material paths have been argued as critical for interpretation of thermochronological data (e.g., Batt and Brandon, 2002; Batt et al., 2004). Figure 10.5 illustrates the approximate spatial variation in denudation rates across the central region of this

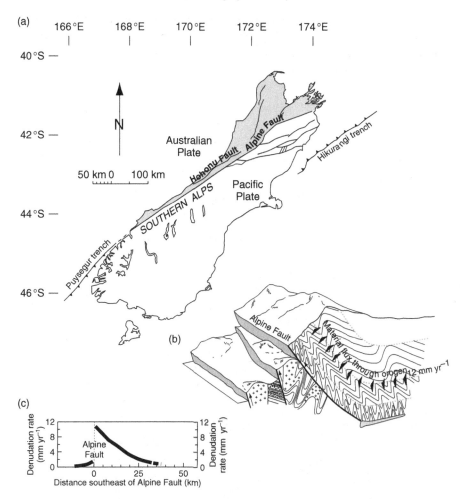

Fig. 10.5. (a) Major geological features of the Australia–Pacific plate boundary through the South Island of New Zealand. The Australian Plate component of the system is shaded light grey, with the Pacific Plate unshaded. (b) A block diagram summarising the structural behaviour of the Plate Boundary system and major fault structures through the region. This is a composite diagram, drawing on the results of Reyners and Cowan (1993), Smith et al. (1995) and Kamp et al. (1992). (c) Approximate exhumation rates across the central Southern Alps, calculated by drawing together the fission-track-dating results of Kamp et al. (1992) and Tippett and Kamp (1993). Rates are plotted relative to the Alpine Fault.

orogen, in the vicinity of Mount Cook (Tippett and Kamp, 1993), together with supplementary kinematic data.

(i) Calculate the significance of lateral integration (η value) for the Mount Cook region for zircon fission-track ages (which have a local minimum of ≈ 0.5 Myr), biotite ^{40}Ar–^{39}Ar ages (local minimum ≈ 1 Myr) and muscovite ^{40}Ar–^{39}Ar ages (local minimum ≈ 1.5 Myr), and the initiation of active exhumation for material currently exposed in the region, estimated by Tippett and Kamp (1993) as 6.7 ± 0.6 Myr.

(ii) On the basis of these results, what tectonic significance could be ascribed to variation in the average exhumation rates inferred from these ages?

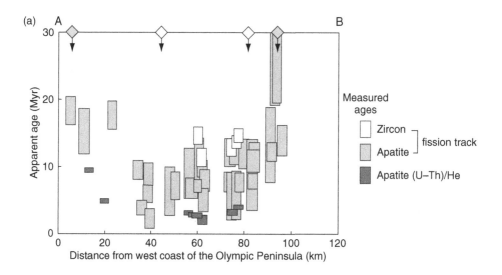

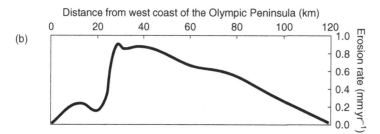

Fig. 10.6. (a) Thermochronological age variation across the Olympic orogen, plotted as apparent age versus distance from the Pacific Coast along a section through the most deeply exhumed core of the orogen, parallel to the convergence vector of the Juan de Fuca Plate. The zircon fission-track ages come from Brandon and Vance (1992) and Garver and Brandon (1994), the apatite fission-track ages from Brandon et al. (1998) and Roden-Tice (unpublished data), and the (U–Th)/He ages are taken from Batt et al. (2001) and unpublished data of the authors. The shaded symbols straddling the upper margin of the chart indicate the transition to inherited, or 'unreset' ages off the scale of this figure for the respective fission-track datasets. (b) Recent erosion rates across the Olympic Peninsula, calculated from apatite fission-track ages and deformed and offset river terraces, after Pazzaglia and Brandon (2001).

10.5 Evaluation of the significance of lateral variation

Tutorial 10

As discussed in Section 10.5 above, the kinematic synthesis of the Cascadia Forearc Wedge and Olympic Mountains favoured by Batt *et al.* (2001) combines to yield η values of 0.27 for apatite (U–Th)/He ages, 0.67 for apatite fission-track ages and 1.73 for zircon fission-track ages.

(i) How does this comparison help to explain the patterns of variation in thermochronological age across the Olympics relative to the denudation rates derived by Pazzaglia and Brandon (2001) (Figure 10.6)?

(ii) What would the relative age distributions be under the alternative hypothesis of underplating beneath the Olympic Mountains and vertical exhumation without significant lateral motion considered by Brandon *et al.* (1998) and Batt *et al.* (2001)?

11

Isostatic response to denudation

This chapter investigates the effect of isostatic rebound caused by surface erosion on the distribution of thermochronological ages at the surface of the Earth. We demonstrate that age–elevation relationships contain information on the nature of this isostatic response, i.e. on the degree of flexural compensation, and can, in effect, be used to provide constraints on the effective elastic thickness of the continental lithosphere.

11.1 Local isostasy

The post-orogenic phase of most mountain belts is characterised by a gradual erosion of the topography created during the active tectonic phase. This erosion results in unloading of the underlying lithosphere and consequent isostatic adjustment. The principle of isostasy assumes that there is a region beneath the lithosphere (the *compensation depth*) where rocks are so weak that they cannot sustain any horizontal stress gradient over geological times, and hence the region adjusts to imposed loads by deformation. This implies that, at isostatic equilibrium, the weight of adjacent lithospheric columns must be equal. Erosion of surface topography results in a local reduction of the weight of the underlying lithospheric column and must therefore be compensated by vertical uplift.

As shown in Figure 11.1, surface erosion by an amount E_0 (panel (b)) of a reference lithospheric column (panel (a)) of crustal thickness h_c and total thickness $h_c + L$ results in an isostatic surface uplift by an amount u (panel (c)) such that the weights of the two columns ((a) and (c)) down to the compensation depth $h_c + L$ are identical. This leads to the following relationship:

$$h_c \rho_c + L\rho_m = (h_c - E_0)\rho_c + L\rho_m + u\rho_a \qquad (11.1)$$

11.1 Local isostasy

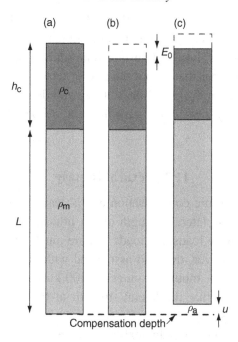

Fig. 11.1. Erosion at the surface of a reference lithospheric column (a) by an amount E_0 (b) leads to isostatic uplift of the column by an amount u (c). For typical values of the density and crustal thickness, $u \approx 0.8 E_0$.

where ρ_c, ρ_m and ρ_a are the crustal, lithospheric mantle and asthenospheric densities, respectively, from which an expression for the isostatic uplift u as a function of the erosion, E_0, can be derived:

$$u = E_0 \frac{\rho_c}{\rho_a} \qquad (11.2)$$

From this relationship, one can also derive the expression for the amount of erosion, E, needed to reduce the topography by an amount Δh:

$$E = \frac{\Delta h}{1 - \rho_c/\rho_a} \qquad (11.3)$$

which, for average values of upper-crustal and asthenospheric densities of 2600 and 3200 kg m^{-3}, respectively, leads to an isostatic amplification factor of approximately 5.3. This means that, for every kilometre of surface topography reduction, approximately 5.3 km of crustal rocks must be removed by erosional or structural processes. This also means that, during the post-orogenic phase of a mountain belt, erosion of the topography may lead to tens of kilometres of exhumation and the subsequent resetting of thermochronological systems, especially those characterised by low closure temperatures, such as the (U–Th)/He and fission-track systems in apatite. Consequently, for those thermochronological systems, ages observed at the surface of an ancient mountain belt may provide information about the rate of erosion and evolution of surface relief during the post-orogenic phase of the mountain belt rather than about the rate of tectonic uplift and exhumation during its constructive (or orogenic) phase (e.g., van der Beek and Braun, 1998), and need to be carefully interpreted accordingly to deconvolve these signals.

11.2 Flexural isostasy

The concept of an 'isostatic compensation depth' leads to a vertically integrated mass balance only if the lateral strength of the lithosphere is neglected. This is usually true for very wide loads, i.e. loads that are much wider than the thickness of the lithosphere. However, the load associated with erosional processes at the scale of a single valley or mountain range (1–100 km) is typically less than this limiting condition, such that the lateral strength of the lithosphere can support some proportion of this load.

The lateral strength of the continental lithosphere is usually taken into account by parameterising its isostatic response as that of a thin, yet strong, elastic plate 'floating' on the underlying inviscid asthenosphere (Turcotte, 1979). Under the assumption that the surface deflection of the lithosphere, w, is small in comparison with its thickness, one can derive the following partial differential equation that relates w to an applied vertical surface load, $q(x)$:

$$D \frac{\partial^4 w}{\partial x^4} + (\rho_a - \rho_s)gw = q(x) \qquad (11.4)$$

where D is the flexural rigidity of the plate, ρ_a is the density of the asthenosphere and g is the acceleration due to gravity (Turcotte, 1979). What ρ_s represents depends on the assumption made regarding the evolution of the surface following its deflection by elastic rebound; if subsidence is accompanied by sedimentation, ρ_s is the sediment density; if no sedimentation takes place, ρ_s is the density of water; where the surface is uplifted but no erosion takes place, ρ_s should be the upper-crustal rock density; where the surface is uplifted and erosion is very efficient, ρ_s should be zero.

The flexural rigidity can be expressed in terms of the elastic constants, Y_m and ν (Young's modulus and Poisson's ratio), and the assumed equivalent elastic thickness of the plate, T_e (Turcotte, 1979):

$$D = \frac{Y_m T_e^3}{12(1 - \nu^2)} \qquad (11.5)$$

11.3 Periodic loading

The flexural isostatic response of the lithosphere to a periodic surface load (resulting, for example, from a periodic surface topography of amplitude h_0) can be obtained from the solution of Equation (11.4) (Turcotte and Schubert, 1982), in which

$$q(x) = \rho_c g h_0 \sin\left(\frac{2\pi x}{\omega}\right) \qquad (11.6)$$

The solution is in phase with the load:

$$w = w_0 \sin\left(\frac{2\pi x}{\omega}\right) \qquad (11.7)$$

and its amplitude, w_0, is given by

$$w_0 = \frac{h_0}{\dfrac{\rho_a}{\rho_c} - 1 + \dfrac{D}{\rho_c g}\left(\dfrac{2\pi}{\omega}\right)^4} \qquad (11.8)$$

The amplitude of the deformation depends on the wavelength of the load, ω, relative to the flexural wavelength, x_α, which is given by

$$x_\alpha = \left(\frac{D}{\rho_c g}\right)^{1/4} \qquad (11.9)$$

Two end-member cases exist. Firstly, when the wavelength of the topography is small in comparison with the flexural wavelength,

$$\omega \ll x_\alpha \qquad (11.10)$$

the deflection of the plate, and thus the isostatic response of the lithosphere, becomes negligible in comparison with the amplitude of the surface topography:

$$w_0 \ll h_0 \qquad (11.11)$$

This means that the load of the topography is small enough to be fully compensated by the flexural strength of the lithosphere. Secondly, in cases where the wavelength of the topography is much greater than the flexural wavelength,

$$\omega \gg x_\alpha \qquad (11.12)$$

the flexural strength of the lithosphere becomes negligible and the topography is in local isostatic equilibrium:

$$w_0 = \frac{\rho_c h_0}{\rho_a - \rho_c} = w_0^e \qquad (11.13)$$

The degree of isostatic compensation, C, of a topographic load of wavelength ω is defined as the ratio of the deflection of the lithosphere to its maximum local isostatic (or hydrostatic) deflection (Turcotte and Schubert, 1982):

$$C = \frac{w_0}{w_0^e} = \frac{\rho_a - \rho_c}{\rho_a - \rho_c + \frac{D}{g}\left(\frac{2\pi}{\omega}\right)^4} \tag{11.14}$$

For small-wavelength topography ($\omega \ll x_\alpha$), C tends towards 0; for long-wavelength topography ($\omega \gg x_\alpha$), C tends towards 1.

11.4 Isostatic response to relief reduction

Whether topography is in isostatic equilibrium or fully compensated by the lateral strength of the lithosphere will affect the amount of isostatic amplification associated with post-orogenic relief reduction by erosion. To follow the convention used earlier (Section 8.1), the modern topographic amplitude is assumed to be a multiple β of what it was at the end of the active orogenic phase (i.e. before the post-orogenic decay started). From Equation (11.8), one can deduce the total erosion, E, associated with the reduction in topographic amplitude as the sum of two terms, one representing the lowering of the surface topography, E_1, the other representing the isostatic uplift in response to the erosional unloading, E_2:

$$E = E_1 + E_2 = \left(\frac{1}{\beta} - 1\right)h_0 + \left(\frac{1}{\beta} - 1\right)h_0 \frac{\rho_c}{\rho_a - \rho_c} C$$
$$= \left(\frac{1}{\beta} - 1\right)h_0 \left(1 + \frac{\rho_c}{\rho_a - \rho_c} C\right) \tag{11.15}$$

Equation (11.15) is the generalization of Equation (11.3) that gives the total amount of erosion, E, necessary to change the surface topographic amplitude by a factor $1/\beta$ for a topography characterised by a wavelength ω, on the basis of the assumption that the lithosphere has a flexural strength that can be represented by the deflection of a thin elastic plate.

11.5 Effects on age distribution

The effect of isostatic uplift on the distribution of thermochronological ages is demonstrated in the study of Braun and Robert (2006). The authors apply a version of the Pecube software modified by the inclusion of a module in which the spectral method described in Nunn and Aires (1988) is used to calculate the vertical deflection of a thin elastic plate due to a vertical load applied at the surface. Flexural isostasy is incorporated by computing the negative load and

11.5 Effects on age distribution

isostatic uplift resulting from the imposed change in surface topography, Δh, at each time step. This uplift is then imposed as a velocity term in a system of reference that is fixed with respect to the base of the model.

Rather than using a synthetic topographic surface, Braun and Robert (2006) used a topographic dataset from the Dabie Shan area in southeastern China. The reason for this choice was that there already existed an extensive thermochronological dataset for this area (Reiners *et al.*, 2003b). This small orogen developed between the northern edge of the Yangtze craton and the southeastern corner of the Sino-Korean craton during a series of subduction-related episodes of crustal shortening, from the late Palaeozoic to the mid-Cretaceous (Schmid *et al.*, 2001). There is debate on whether the orogen was reactivated during the ongoing Indo-Asian collision and whether some of the present-day topographic relief is the result of this Cenozoic reactivation (Grimmer *et al.*, 2002). Alternatively, the topography may be the erosional remnant of a much larger amplitude relief that was entirely formed during the Cretaceous (Reiners *et al.*, 2003b).

Using the topography of this area (extracted from the 30-arc-second-resolution DEM GTOPO30; see the topographic profile in Figure 11.2) Braun and Robert (2006) performed a series of model runs in which they systematically varied the effective elastic thickness of the lithosphere. The topographic relief was arbitrarily

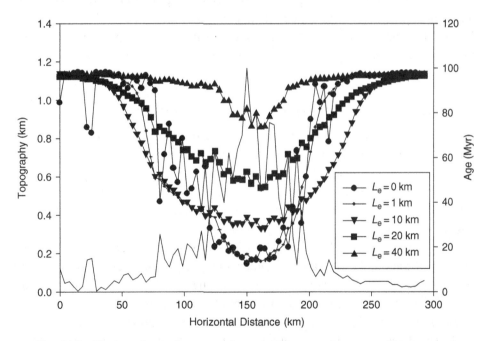

Fig. 11.2. Topography and computed apatite-He ages along a profile extracted from a DEM of the Dabie Shan area (from the NW to the SE), assuming various values for the thickness of elastic plate.

imposed to be four times larger 100 Myr ago than it is today and allowed to decay linearly since then, to reach its present-day value at the end of the model run. The results of five model experiments are shown in Figure 11.2 as five predicted apatite (U–Th)/He age distributions across the strike of the orogen, from the northwest to the southeast. The model runs differ by the assumed thickness of elastic plate, 0, 1, 10, 20 and 40 km, respectively, which correspond to flexural wavelengths of 0, 4.1, 23, 39 and 66 km, if one assumes that $Y_m = 10^{11}$ Pa, $\nu = 0.25$ and $\rho_c g = 3 \times 10^4$ Pa m^{-1}. The apatite-He ages are computed from temperature–time histories derived from the results of the Pecube model using a method similar to that described in Section 2.5.

In all model runs, ages are low near the centre of the orogen (Figure 11.2), i.e. they are smaller than the duration of the model run (100 Myr). The minimum reset ages, which are found near the centre of the orogen, are inversely proportional to the assumed thickness of elastic plate. Because it is assumed that the rate of decrease in surface topographic relief is constant through time, the larger the total erosion, the higher the specific rates of erosion and exhumation experienced, and the younger the consequent thermochronological ages exposed at the surface. For low values of the elastic-plate thickness (0–1 km), the system is at or near local isostatic equilibrium and the reduction in surface topographic relief over the last 100 Myr causes very large (up to 6 km) isostatic rebound and erosion. As the elastic-plate thickness is increased, the amount of isostatic rebound decreases and the total amount of erosion necessary to reduce the surface topographic relief by the imposed factor of four decreases accordingly.

11.6 Effects on age–elevation distributions

On the scale of the orogen, the ages produced by this isostatic rebound are inversely proportional to (present-day) elevation, i.e. ages are younger near the centre of the orogen where the topography is currently the largest. However, on the scale of an individual valley (10 km length scale), three distinct styles of relationship between age and elevation are observed under different conditions (Figure 11.2). In cases for which the flexural wavelength is larger than the width of the valleys ($T_e = 10, 20, 40$ km), there is a strong positive correlation between age and elevation; in the case for which the flexural wavelength is similar to the width of the smallest valleys ($T_e = 1$ km), there is little variation in age with elevation; in the case for which the flexural wavelength is smaller than the valley width ($T_e = 0$ km), the predicted age is inversely proportional to the elevation.

The small-scale topographic features are characterised by a wavelength (10 km) that is larger than the critical wavelength, λ_c, given by the ratio of the closure temperature for the thermochronological system considered (75 °C) and the geothermal gradient (20 °C km^{-1}) (see Section 8.1). The perturbation of the closure temperature isotherm caused by the valleys is therefore moderate and the slope of the age–elevation profiles should provide a reasonably accurate, yet overestimated, measure of the local denudation rate (cf. Sections 8.1, 6.2 and 6.3). In the case for which the elastic thickness is large ($T_e = 20$ or 40 km) and isostatic rebound is negligible, the total erosion is equal to the change in topographic amplitude. When isostatic rebound becomes important (i.e. for smaller values of the assumed elastic-plate thickness), the total erosion and, consequently, the mean erosion rate increase. This is why the model predicts that the apparent erosion rate derived from the slope of the age–elevation relationship measured along narrow valley profiles increases with decreasing elastic-plate thickness.

11.7 Application to the Dabie Shan

Rocks were collected across the Dabie Shan and dated by (U–Th)/He and apatite and zircon fission-track methods (Reiners *et al.*, 2003b). In their study, Braun and Robert (2006) focussed only on the low-temperature, apatite datasets (Figure 11.3). In both datasets, there is a general trend of younger ages near the core of the orogen and older ages around its rim. An age–elevation transect collected near the centre of the orogen yields well-defined age–elevation relationships for the He dates,

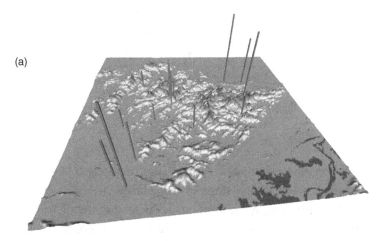

Fig. 11.3. Relationships among age, topography and location for the apatite-He dataset (a) and apatite-fission-track (FT) dataset (b) collected by Reiners *et al.* (2003b) in the Dabie Shan. Each bar corresponds to an age measurement. The location of the bar gives the location of the sample; the height of the bar is proportional to the measured age. The maximum age for apatite-He measurements is 65.5 Myr; the maximum age for apatite-FT measurements is 85.7 Myr.

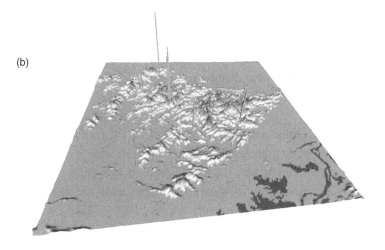

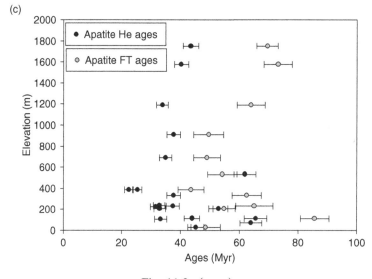

Fig. 11.3. (cont.)

with a slope, or apparent exhumation rate, of approximately $0.064\,\mathrm{km\,Myr^{-1}}$, and for the fission-track dates, with a slope of $0.042\,\mathrm{km\,Myr^{-1}}$.

It has been demonstrated in the previous sections that the distribution of ages within the orogen (i.e. as a function of the distance to the orogen's centre) and the relationship between age and elevation significantly vary as the assumed thickness of elastic plate is varied. As shown in Braun and Robert (2006), the numerical model Pecube can be used not only to search through parameter space for the 'optimal' set of parameters that result in age predictions similar to the observed distribution of ages (within measurement error), but also to evaluate the sensitivity

11.7 Application to the Dabie Shan

of the model predictions to the values of the input parameters (see Section 8.5 for a general discussion on inversion methods).

To perform this inversion, Braun and Robert (2006) used the Neighbourhood Algorithm of Sambridge (1999a), which we briefly described in Section 8.5. The misfit function is defined as the L_2-norm of the weighted difference between the observation vector, **O**, and the prediction vector, **P**:

$$\text{misfit} = \frac{1}{n}\sqrt{\sum_i^n \left(\frac{O_i - P_i}{\Delta O_i}\right)^2} \qquad (11.16)$$

where n is the number of measured ages (31 in this case study) and ΔO_i are the observational errors. The 'free' model parameters, i.e. those for which the inversion is performed, were selected to be the effective elastic-plate thickness, T_e, the length of the model run, t_e, which can also be regarded as the time since the last major tectonic event at the end of which the erosional decay episode started, the basal temperature, T_1, the amplitude of the decrease in topography/relief since the end of the orogenic phase, β^{-1}, and an additional mean exhumation rate, $\dot{E}$,

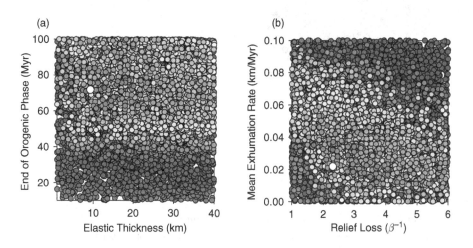

Fig. 11.4. Results of the Neighbourhood Algorithm inversion as scatter diagrams of the misfit between observations and predictions. Each circle corresponds to a forward model run. The position of the circle is determined by the values of the model parameters. The colour of the circle is proportional to the value of the calculated misfit. Red corresponds to low misfit values; blue corresponds to high misfit values. The larger, white circle corresponds to the best-fit model run. Each diagram corresponds to a projection of all model runs onto a plane defined by two of the five parameters. We show here only a small number (five) of all possible combinations of pairs of parameters.

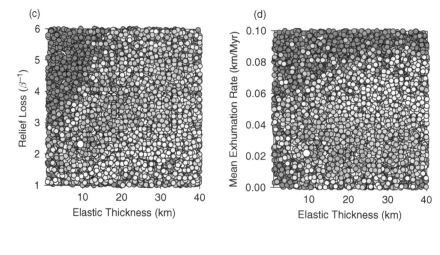
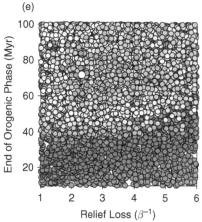

Fig. 11.4. (cont.)

i.e. uniform in space and constant over time, that is independent of the isostatic flexural rebound.

Braun and Robert (2006) used a high-performance computer cluster to perform a very large number of forward model runs (17 744 in total) within a reasonable amount of time (a few days). The results are shown in Figure 11.4 as scatter plots in parameter space. Each circle corresponds to a Pecube forward model run. The shade of the circles is proportional to the predicted misfit (dark greys correspond to low misfit; light greys correspond to high misfit).

The results demonstrate that the thermochronological data display sensitivity to the timing of the end of the orogenic phase (60 Myr $< t_e <$ 80 Myr), the effective elastic-plate thickness ($T_e <$ 20 km), the amplitude of the relief loss ($2 < \beta^{-1} < 4$) and the mean exhumation rate (0.01 km Myr$^{-1} < \dot{E} <$ 0.04 km Myr^{-1}). The data are less sensitive to the basal temperature, T_1. The best-fitting forward model run (indicated on Figure 11.4 by a large white circle) is obtained with the following parameter values: $T_e = 9.3\,\text{km}$, $t_e = 72\,\text{Myr}$, $T_1 = 660\,°C$, $\beta^{-1} = 2.3$ and $\dot{E} = 0.022\,\text{km}\,\text{Myr}^{-1}$. The ages predicted from this best-fitting model run are shown and compared with the observed ages in Figure 11.5 as age–elevation distributions for the He and fission-track ages. The best-fit forward model run produces age distributions mimicking the steep, positive correlation between age and elevation

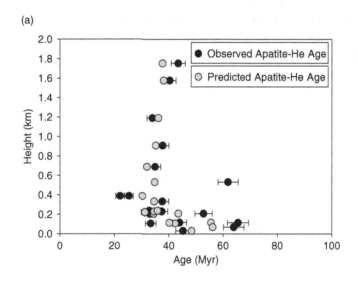

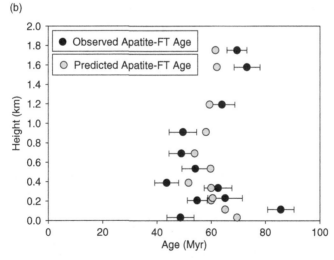

Fig. 11.5. Comparison between observed and synthetic age–elevation relationships for apatite-He ages (a) and apatite-fission-track (FT) ages (b) estimated from the best-fitting model run.

near the centre of the orogen and the increase in age with distance from the centre of the orogen (and thus mean elevation). For the parameter values of the 'best-fit model' the total erosion in the centre of the orogen is approximately 4 km, i.e. 1.6 km of uniform erosion, 2 km of relief loss and 0.4 km of associated isostatic rebound.

This example demonstrates that, when comprehensively analysed, the information contained in thermochronological datasets can aid in the constraint of a much wider range of Earth processes than the approach is commonly applied to. Not only can relief evolution, the timing of orogenic events and the mean exhumation rate be extracted from the distribution of cooling ages in surface rocks, but also a strong link to the mechanical response of the lithosphere to surface loading can be demonstrated by analysis of the relationship between age and elevation in a carefully sampled dataset.

12

The evolution of passive-margin escarpments

In this chapter, we will apply several techniques described in the earlier sections to derive constraints on the evolution of continental passive-margin escarpments from low-temperature thermochronological datasets. In doing so, we will show how a three-dimensional finite-element solver of the heat-transport equation (Pecube) can be coupled to the predictions of a landscape-evolution model (Cascade) to demonstrate the sensitivity of thermochronological data to various geomorphic scenarios. We use an inverse method to demonstrate what can and, potentially most importantly, what cannot be constrained from a given thermochronological dataset. We will also show how numerical modelling can be used to devise efficient and meaningful data-collection strategies.

12.1 Introduction

Great escarpments along high-elevation rifted continental margins form some of the most prominent morphological features on Earth (Figure 12.1). Since the early 1990s, the geoscience community has re-examined these features through an array of quantitative processes, leading to an increasing appreciation of the contribution of escarpment evolution to the dynamics of rifted margins (e.g., Beaumont *et al.*, 1999; Gilchrist and Summerfield, 1990, 1994; van der Beek *et al.*, 1995). At the same time, the geomorphological community has shown a renewed interest in large-scale, long-term landscape development of rifted margins and other intra-plate settings (e.g., Summerfield, 2000). This work is driven by an array of fundamental questions surrounding the formation and development of high-elevation rifted margins. What is the tectonic significance of rift-flank escarpments? How do they evolve, and how can we interrogate their geological record to derive constraints on the rift systems to which they relate?

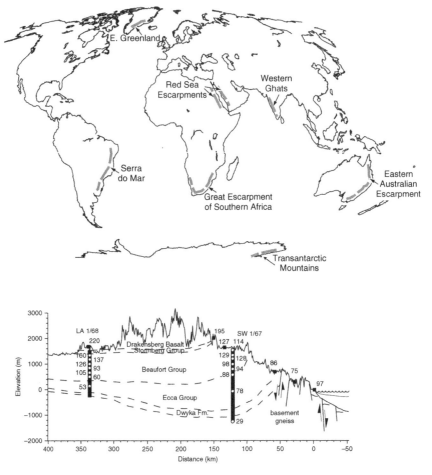

Fig. 12.1. Locations of major escarpment systems at rifted continental margins discussed in this chapter. Modified from Kooi and Beaumont (1994). The lower panel shows a topographic and simplified geological cross-section across the southeastern African (Drakensberg) escarpment. Numbers are apatite-fission-track ages from surface samples as well as from two drill-cores (after Brown et al. (2002)). Reproduced with permission from the American Geophysical Union.

Rifted margins are characterised by amounts of denudation of the order of a few kilometres, which means that apatite fission-track and (U–Th)/He data will typically record the full history of their development, whereas in orogenic regions (e.g., Chapter 13) they may record only the last few Myr in the evolution of the belt. Over the past 20 years, the development of apatite fission-track thermochronology (e.g., Bohannon et al., 1989; Brown et al., 1990; Gallagher et al., 1994; Gunnell et al., 2003) and, more recently, apatite (U–Th)/He (Persano et al., 2002) and cosmogenic isotope analysis (Bierman and Caffee, 2001; Cockburn et al., 2000; Fleming et al., 1999; van der Wateren and Dunai, 2001) has contributed significantly to quantifying the denudational history of rifted margins. These data are in many cases incompatible with traditional paradigms for the evolution of such margins, which explained the observed morphology in terms of pulses of uplift and escarpment retreat (e.g., King, 1962; Ollier, 1985). The new data and models have led to the realisation that the geomorphic evolution of escarpment systems may be considerably more complex and variable (Brown et al., 2000; Gallagher and Brown, 1997; Gilchrist and Summerfield, 1994) than had previously been thought.

The interaction between tectonics and erosion on rifted margins is relatively simple: available constraints suggest that there may have been a pulse of syn-rift uplift, after which the response to denudation appears to have been purely flexural isostatic. This makes numerical models for landscape development on rifted margins relatively simple, compared with those used for compressional orogens (see, for instance, Beaumont et al. (1999) and Tucker and Slingerland (1994)). This simplicity has led to the use of numerical surface-process models (SPMs), also called landscape-evolution models (LEMs), to study rifted margins from the early stages of development of these models.

12.2 Early conceptual models: erosion cycles

Conceptual models for landscape development on passive continental margins date back to the late nineteenth century, when Suess (1906) and especially Davis (1892) proposed a theoretical framework in which the geomorphic development of rifted margins was to be cast for nearly a century. Within their concept of 'erosion cycles', rifted margins were suggested to conform to a generally applicable model of evolving through pulses of rapid tectonic uplift followed by long periods of erosional downwearing and 'ageing' of the landforms. This would lead to the establishment of planation surfaces that could, it was argued, be dated by correlating them with offshore sequences or by directly dating deposits or alteration products associated with them. During the 1950s, Leister C. King (King, 1955, 1962) adapted the Davisian ideas in such a way as to involve the creation of planation surfaces through escarpment retreat rather than by downwearing.

The paradigm of cyclic landform chronology has proved very influential and long-lived. The concept remains at the foundation of the geomorphic literature that is currently embraced by geodynamicists studying vertical motions through time of regions like southern Africa (e.g., Gurnis et al., 2000). However, although the existence of flat landscape elements that can locally be dated is not debated, the correlation of these remnants and the notion that they were initially horizontal and continuous (two necessary assumptions if one intends to use them to establish uplift and denudation chronologies) are controversial (Summerfield, 2000).

12.3 Thermochronological data from passive margins

The first fission-track thermochronology studies of passive continental margins date from the mid 1980s and their authors focussed on the timing of margin uplift with respect to rifting, hoping to use the results as an argument in the debate on 'active versus passive rifting' that had dominated thinking since the 1970s (Sengör and Burke, 1978). Results from these first studies in SE Australia (Moore *et al.*, 1986) and along the Red Sea (Bohannon *et al.*, 1989; Omar *et al.*, 1989) showed that apatite fission-track ages were generally equal to or younger than rifting on the coastal strip seaward of the escarpment, and much older inland (Figure 12.2). This indicated that several kilometres of denudation (and, it was argued at that time, uplift) had taken place since the onset of rifting on these margins, and therefore favoured a 'passive' rifting mechanism. The fission-track data also clearly showed that the patterns of denudation at rifted margins are incompatible with the conceptual model of a 'rift-margin monocline' that was based on the correlation of planation surfaces inland and outboard of the escarpment with offshore sedimentary-sequence boundaries (e.g., Gallagher *et al.*, 1998; Brown *et al.*, 2000).

However, the fact that fission-track ages seaward of the escarpment are younger than rifting shows only that cooling of that region, which was presumably denudation controlled, post-dates rifting. The mechanism driving denudation of the coastal strip is most probably the increase in local relief imprinted by rifting; significant amounts of denudation may occur without requiring tectonic uplift of the margin at all (Gallagher *et al.*, 1994; van der Beek *et al.*, 1999; Brown *et al.*, 2000) (see Chapter 11). Flexural backstacking of the amount of overburden removed from the margin may indicate whether it has undergone rift-flank uplift (e.g., van der Beek *et al.*, 1994) but inferences from such an exercise depend strongly on the flexural model that is chosen.

In some regions, the amounts of denudation inferred from the fission-track data have been in disaccord with the rates of landscape change inferred from geomorphological studies. This point has notably raised a 10-year long controversy in SE Australia (see Kohn and Bishop (1999) for a review). The amounts of denudation inferred from apatite fission-track data obviously depend on the geothermal gradient, which is generally very poorly constrained for onshore passive margins. Alternatively, the spatial and temporal scales to which the fission-track and the geomorphological studies pertain are often not the same, and the disagreement may be a result of unwarranted extrapolation of results rather than reflecting a real discrepancy (van der Beek *et al.*, 1999).

Results of early studies suggested that the fission-track data could actually record syn-rift heating and post-rift cooling of the margin, combined with limited denudation (e.g., Moore *et al.*, 1986; Dumitru *et al.*, 1991; O'Sullivan *et al.*,

12.3 Thermochronological data from passive margins

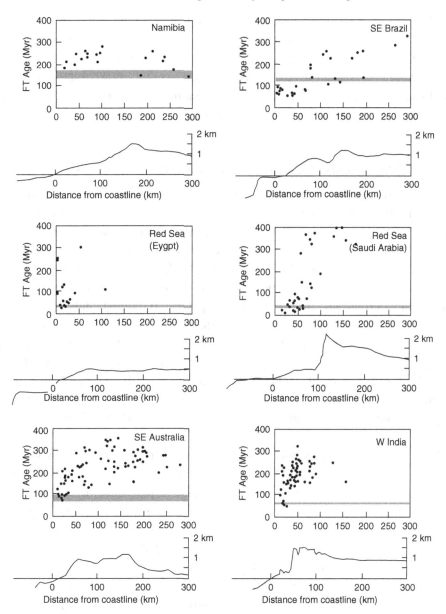

Fig. 12.2. Topographic profiles and apatite-fission-track (FT) ages across six passive continental margins. The grey bars in the FT age plots indicate the timing of break-up. Modified from Gunnell (2000). Reproduced with permission from Blackwell Publishing.

1995). Several numerical analyses have, however, indicated the impossibility of sufficiently heating the upper few kilometres of crust in a rifted margin by lithospheric thinning or even magmatic underplating (Gallagher *et al.*, 1994; van der Beek, 1995). In contrast, the nature of the overburden and its thermal conductivity, as well as the possibility that fluid-flow systems have operated in the past, may strongly control the evolution of the near-surface thermal structure of onshore passive margins (e.g., Gallagher *et al.*, 1994). In southeastern Australia, for instance, the youngest apatite fission-track ages of ≤100 Myr occur nearly exclusively in areas formerly covered by sediments and are therefore possibly affected by blanketing effects (cf. Section 4.8).

12.4 Models of landscape development at passive margins

Numerical modelling of landscape development on rifted margins can dramatically improve understanding and interpretation of thermochronological data in these settings. Few prior models, however, have explicitly attempted to explain the detailed spatial distribution of thermochronological ages on passive continental margins. Early on, two-dimensional models demonstrated the importance of denudation and resulting isostatic rebound in generating, maintaining and modifying the morphology of rifted-margin upwarps (Gilchrist and Summerfield, 1990; ten Brink and Stern, 1992; van der Beek *et al.*, 1995). More sophisticated, planform, SPMs have been employed to investigate both the conditions that are necessary to generate and maintain escarpments on high-elevation rifted margins (Kooi and Beaumont, 1994; Tucker and Slingerland, 1994) and the controls exerted by factors such as lithology and pre-break-up morphology on the subsequent evolution of such margins (Gilchrist *et al.*, 1994; Kooi and Beaumont, 1994, 1996).

Numerical landscape-evolution models contain three main components (Figure 12.3): the initial and boundary conditions imposed by tectonics, a number of surface-process algorithms and a tectonic/isostatic response to erosion. The models that have hitherto been applied to rifted margins take a simple kinematic approach to the tectonic control; i.e., the pre-rift morphology and syn-rift uplift history are imposed on the model. Surface processes incorporated in the models usually are fluvial incision and transport, hillslope transport and bedrock landsliding. These are cast in terms of numerical algorithms of varying complexity: a diffusion law is commonly used for hillslope transport; various mechanical and stochastic models exist for landsliding (e.g., Densmore *et al.*, 1998); and fluvial transport is often modelled using what is known as the 'stream power' law (Whipple and Tucker, 1999). These laws are allowed to act on the initial landscape for a predefined period of time, and the isostatic response of the lithosphere to denudation may be included using flexural models. Predictions of landscape-evolution models include the present-day, end morphology and drainage pattern but also the denudation history at every point in the model, from which thermochronological ages can be predicted if the thermal structure of the crust is included.

12.4 Models of landscape development at passive margins

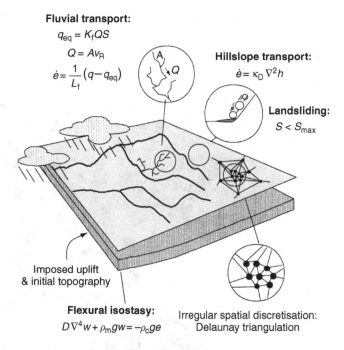

Fig. 12.3. A cartoon of the Cascade surface-processes model. Modified from van der Beek *et al.* (1999, 2001). Reproduced with permission from Blackwell Publishing and the University of Chicago Press.

An example of how apatite fission-track data and landscape evolution models can be combined to study the geomorphic development of rifted continental margins is provided by a recent study of the eastern margin of South Africa (the Drakensberg Escarpment). The southeastern African margin was formed by oblique opening of the Natal Basin about 130 Myr ago. The morphology of southeast Africa is characteristic of a high-elevation rifted margin, with a prominent erosional escarpment (the Drakensberg Escarpment) separating the high-standing continental interior (the Lesotho Highlands) from a strongly dissected coastal region (Figure 12.4).

A fission-track database including samples from two boreholes (Brown *et al.*, 2002) has shown that some 4.5 km of overburden has been removed from the coastal strip of this margin since break-up, with denudation rates peaking in the early–mid Cretaceous. Samples from a borehole located 50 km seaward of the Drakensberg Escarpment record a phase of accelerated denudation from ∼90–70 Myr ago, much earlier than would be expected in the case of an escarpment retreating at a constant rate since break-up. Brown *et al.* (2002) therefore proposed a model of landscape development in which any escarpment initiated at the coast was rapidly destroyed by rivers draining from an interior divide just

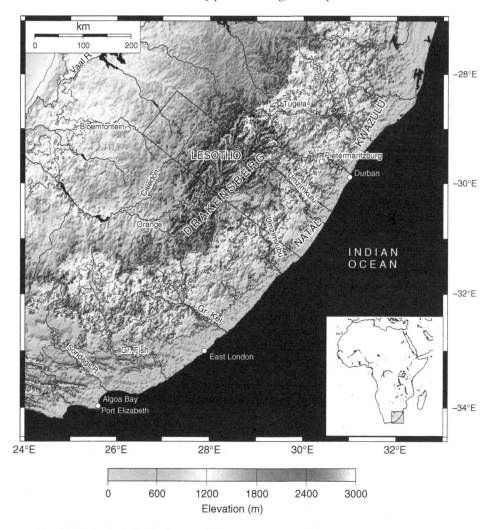

Fig. 12.4. A shaded relief map of the southeastern African margin, colour coded according to elevation and showing localities referred to in the text. The inset map shows the location of the study area on the African continent. The box indicates the location of the modelled cross-section (cf. the inset in Figure 12.1). After van der Beek et al. (2002). Reproduced with permission from the American Geophysical Union.

seaward of the present location of the Drakensberg Escarpment. The present-day escarpment developed and became pinned at this divide.

Van der Beek et al. (2002) designed a numerical model to simulate this behaviour and contribute to understanding of the major controls on landscape development at this margin. The model results show that the pre-break-up topography of the margin has exerted a fundamental control on subsequent margin evolution (Figure 12.5): if the initial topography is a horizontal plateau, the margin

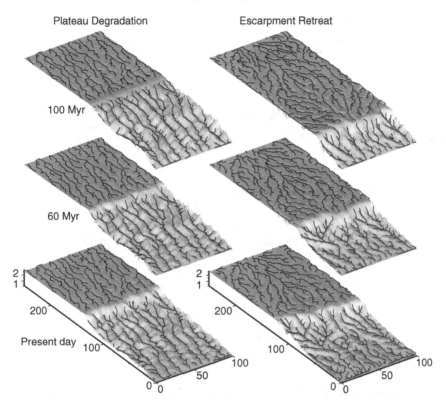

Fig. 12.5. Three-dimensional plots of topography and drainage patterns predicted by the plateau degradation and escarpment-retreat models, at 100 Myr ago, 60 Myr ago and the present day. The models were designed to simulate the evolution of the SE African (Drakensberg Escarpment) margin. After van der Beek *et al.* (2002). Reproduced with permission from the American Geophysical Union.

evolves through escarpment retreat; if the pre-rift topography included an inland drainage divide, the initial plateau seaward of this divide will be rapidly degraded and a new escarpment develops at the location of the initial divide. Although the present-day morphologies predicted by these two models are quite similar, the denudation histories at any point of the margin are very different (Figure 12.6) and it should, in principle, be possible to discriminate between them using thermochronological data. In the SE African case, the data appear to favour the plateau-degradation scenario. Moreover, the models suggest that no large-scale uplift is required in order to explain the present-day morphology and denudation history of the margin: passive denudation with associated flexural isostatic rebound of a pre-existing ∼2.5-km-high plateau is sufficient. These model results are in agreement with previous simulations of the SW African margin (Gilchrist

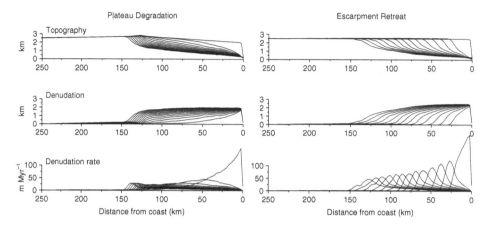

Fig. 12.6. Evolution of strike-averaged topography, denudation and denudation rates for the models shown in Figure 12.5. Profiles are shown at 10 Myr intervals, from zero (start of the model run) to 130 Myr (the present day). After van der Beek et al. (2002). Reproduced with permission from the American Geophysical Union.

et al., 1994) and SE Australia (van der Beek and Braun, 1999), which did not explicitly incorporate the thermochronological data.

Although some first-order conclusions can be drawn from such a qualitative comparison between thermochronological data and model predictions, there remain discrepancies between the inferences from the fission-track data and the outcomes of the geomorphic models. In general, the landscape-evolution models have difficulty adequately reproducing the amounts of denudation inferred from the fission-track data; they also struggle to reproduce the phases of accelerated denudation that are suggested by inverse modelling of the track-length distributions. These problems can be attributed both to the imperfections which still characterise the landscape-evolution models and to the interpretation of the thermochronological data. Moreover, thermochronological data across rifted margins necessarily constitute a finite (generally relatively small) number of samples with associated errors. Do such data in fact constrain geomorphic scenarios?

12.5 Combining thermochronometers and modelling

Two recent developments are contributing to the resolution of these ambiguities: the combination of apatite fission-track data with (U–Th)/He and/or cosmogenic data; and the development of numerical models that take the transient disturbance of thermal structure into account. For southern Africa, the combination of apatite fission-track and cosmogenic data (Cockburn et al., 2000; Fleming et al., 1999) showed that recent short-term rates of denudation and escarpment retreat

12.5 Combining thermochronometers and modelling

(determined from the cosmogenic data) are an order of magnitude lower than those inferred from the fission-track data and this discrepancy directly inspired the development of numerical models such as that shown in Figures 12.5 and 12.6 in order to try to explain these variations. More recently, results from the first (U–Th)/He study of a rifted margin (Persano *et al.*, 2002) showed that (U–Th)/He ages from the coastal strip of SE Australia are similar to apatite fission-track ages of the same samples, again lending support to models of margin evolution that include a rapid phase of plateau degradation during the first several million years after break-up.

Finally, we can now interpret the thermochronological data using numerical models that fully take into account the disturbance of the thermal structure of the crust by denudation and relief development (cf. Chapter 7). A direct comparison between observed thermochronological ages on rifted margins and model predictions requires that the perturbation of the thermal structure due to changing relief be taken into account. As an example, Figure 12.7 shows (U–Th)/He ages predicted by escarpment-retreat and plateau-degradation models, respectively, similar to those shown in Figures 12.5 and 12.6. The topographic evolution was exported from the Cascade landscape-evolution model into the thermal model Pecube, which was then used to predict (U–Th)/He ages across the margin.

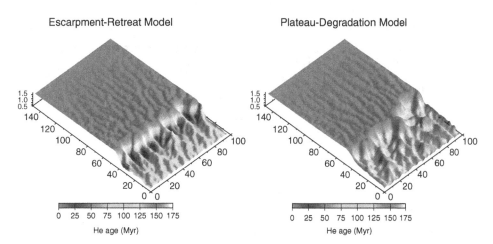

Fig. 12.7. Landforms predicted by Cascade and the apatite (U–Th)/He age distribution predicted from Pecube temperature estimates for the escarpment-retreat model and the plateau-degradation model. The model parameters at the base of this calculation are comparable to those shown in Figures 12.5 and 12.6, except that they were designed to simulate the SE Australian rather than the S. African margin: initial elevation at 1.5 km and rifting since 100 Myr ago. The present-day escarpment is located at ~50 km in both cases. After Braun and van der Beek (2004). Reproduced with permission from the American Geophysical Union.

Note the different patterns of age distribution predicted by the respective models (Figure 12.8): the escarpment-retreat model gives ages clearly decreasing from the coastline to the escarpment, because the peak of denudation migrates inland with the escarpment (e.g., Figure 12.6), whereas the plateau-degradation model predicts less variation in the ages on the coastal strip, which are all close to the age of break-up.

Braun and van der Beek (2004) performed a sensitivity analysis on the boundary conditions, the initial conditions and the parameters of the landscape-evolution model (the initial elevation of the escarpment, retreat rate, equivalent elastic thickness and width of the escarpment zone) in order to test under what circumstances apatite (U–Th)/He and fission-track data can reliably be used to infer the evolution of the margin. To do this, they systematically searched the parameter space, using simplified kinematic models of escarpment development that simulate the behaviour observed in the Cascade models. Conditions that are favourable are those necessary to produce sufficient exhumation to reset the thermochronological system. They include a tall escarpment, a high geothermal gradient and/or a low flexural rigidity of the lithosphere. Braun and van der Beek (2004) demonstrated that, to determine the rate and mode of escarpment migration from low-temperature thermochronology, one needs to collect samples

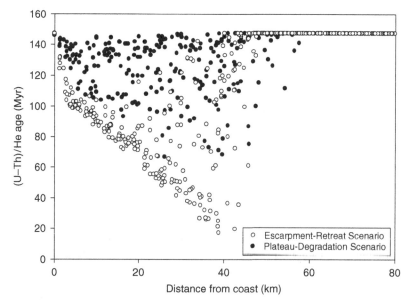

Fig. 12.8. Comparison of the (U–Th)/He ages predicted by the models in Figure 12.7, as a function of distance from the coastline. After Braun and van der Beek (2004). Reproduced with permission from the American Geophysical Union.

12.5 Combining thermochronometers and modelling

along transects perpendicular as well as parallel to the escarpment. The tightest constraints on escarpment development are provided by (in ascending order) the minimum thermochronological age encountered seaward of the escarpment, the location of where the minimum age is found, the slope of the age–distance relationship (in a direction perpendicular to the coast) and the slope of the age–elevation relationship (from a transect parallel to the escarpment).

To address the issue of whether limited and error-prone real-world data constrain evolutionary scenarios, Braun and van der Beek (2004) used the Neighbourhood Algorithm (see Section 8.5) to search the parameter space for best fits to an existing dataset. They used the Bega Valley region of the SE Australian escarpment as a test case, since this is the only area for which both apatite fission-track (e.g., Gleadow et al., 2002; Persano, 2003) and (U–Th)/He (Persano et al., 2002; Persano, 2003) data are currently available. The eastern Australian margin developed as a result of oblique rifting in the Tasman Sea from $\sim$95 Myr ago onwards, with oceanic spreading taking place in the Tasman Sea $\sim$80 Myr ago (Gaina et al., 1998). The margin is characterised by an approximately 1-km-high escarpment running along its entire length of $\sim$2500 km and separating a low-elevation coastal strip from a high-elevation but low-relief upland region. In southernmost New South Wales, where the data were collected (Figure 12.9), the escarpment is cut into Paleozoic granites and metamorphic rocks and is located $\sim$35 km from the coast. In this area, the escarpment forms a secondary drainage divide between linear river systems that flow to the Tasman Sea through relatively deeply incised

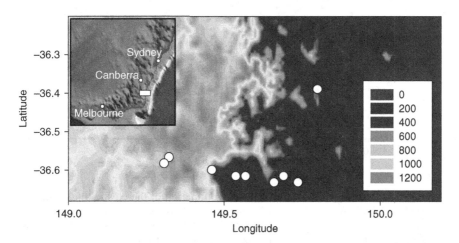

Fig. 12.9. The topography of the study area along the coast of southeastern Australia (rectangle in inset) showing the location of (U–Th)/He samples from Persano et al. (2002). After Braun and van der Beek (2004). Reproduced with permission from the American Geophysical Union.

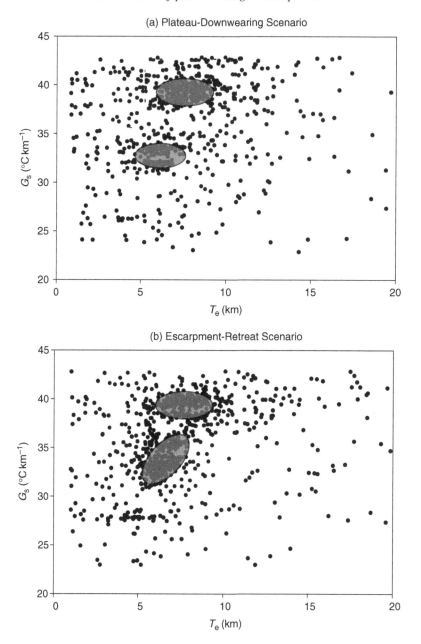

Fig. 12.10. Locations of model runs as a function of the elastic thickness T_e and surface heat flow G_s during a Neighbourhood Algorithm search. As the routine searches the parameter space for best-fit models, the highest density of models (indicated by the grey discs) corresponds to the locations of local minima in misfit. After Braun and van der Beek (2004). Reproduced with permission from the American Geophysical Union.

12.5 Combining thermochronometers and modelling

valleys and the Snowy River and its tributaries on the highland plateau, which flow southwards into the Bass Strait (van der Beek and Braun, 1999).

Whereas (U–Th)/He data are close to the age of break-up (100 Myr) seawards of the escarpment and significantly older inland, apatite fission-track data are older than the age of break-up throughout the area, although they are also young towards the coastline. The inversion results (Figure 12.10) show that the (U–Th)/He data, even though limited, are consistent with low flexural rigidity and/or a high geothermal gradient during and after break-up. They also require escarpment development to have occurred during the first 15 Myr after the onset of rifting, either through downwasting (plateau degradation) or by very rapid retreat, followed by stabilisation at the present-day escarpment location. In this setting, the apatite fission-track data, in contrast, cannot provide this constraint, since rocks have been exhumed from the partial-annealing zone only. This example shows the importance of combining different thermochronological datasets to study rifted-margin evolution and to constrain the numerical models.

13

Thermochronology in active tectonic settings

The coupling between erosion and tectonics is most likely to be efficient in regions of ongoing tectonic activity, especially in regions of continental convergence where the collision between two continents leads to crustal thickening and, by isostasy, to surface uplift. This uplift causes relief (i.e. slope) to be created, which triggers erosion by channel incision or, under colder climatic conditions, the formation of glaciers that will rapidly reshape the landform.

In this chapter, we are going to investigate the effect of continental collision on the temperature structure of the crust; we will also present some of the most widely accepted models for crustal deformation and, most importantly, describe the respective consequences they have for the path followed by rock particles as they travel through the orogen, ultimately to be exposed at the surface. The combination of these processes will provide us with appropriate (that is, quantitative) constraints on the predicted temperature history of particles from which we should be able to predict cooling ages.

At the end of this chapter we will demonstrate how these predictions can then be used to constrain the tectonic evolution of an active mountain belt, including the timing of the onset of convergence, the present-day rate of convergence and its variation along the strike of the orogen. To illustrate this approach, we will consider a dataset collected in the Southern Alps of New Zealand that has been interpreted using state-of-the-art quantitative methods (Batt and Braun, 1999; Herman et al., 2006).

13.1 A simple model for continental collision

In many cases, continental collision can be regarded as the end product of the closure of an oceanic basin bounded on at least one of its sides by a subduction zone (Willett *et al.*, 1993). Sediment that lies on top of the oceanic crust is progressively scraped off by the overlying continent and accumulates in the

13.1 A simple model for continental collision

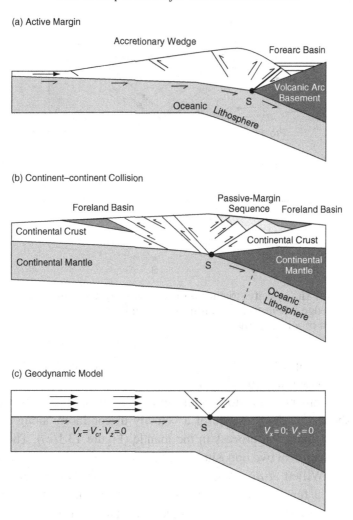

Fig. 13.1. (a) An accretionary prism in a subduction setting. (b) A conceptual model for the collision between two continents in which the mantle part of the lithosphere undergoes subduction while the lighter overlying crust is forced to thicken. (c) The modelling analogue for these two scenarios. Modified from Willett *et al.* (1993). Reproduced with permission from the Geological Society of America.

subduction trough, forming an accretionary prism (Figure 13.1(a)). The internal dynamics of the wedge is adequately represented as that of a *critical wedge* of frictional material at or near failure (Dahlen *et al.*, 1984).

The concept of a critical wedge can be extended to describe the behaviour of the continental crust at a convergent plate boundary in the absence of oceanic lithosphere. Because of their lower density, continental rocks are unlikely to be subducted into the underlying denser mantle. Rather, it can be considered that

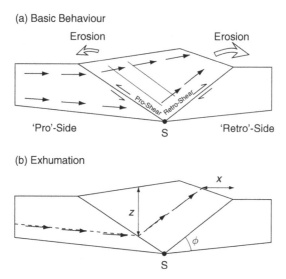

Fig. 13.2. (a) The velocity structure within an actively deforming doubly-vergent orogenic wedge where tectonic influx is in equilibrium with erosional outflux. (b) The path of a rock particle.

they will be decoupled from the underlying continental lithospheric mantle and deformed, as shown in Figure 13.1(b).

During a continental collision of this sort, the crust can be regarded as being forced to shorten by the presence of a velocity discontinuity along its base that represents the subduction process in the mantle (Figure 13.1(c)). The shortening causes the formation of two oppositely dipping shear zones that root in the velocity discontinuity (Willett *et al.*, 1993) (Figure 13.2(a)). The exact geometry of the shear zones is a function of the rheology of the crust; i.e. whether it is mostly brittle or ductile (Willett, 1999b) or whether it is characterised by strain softening (Beaumont *et al.*, 1996). This dynamic model for the evolution of an orogenic belt has been termed the *doubly-vergent critical-wedge model* (Koons, 1990, 1994; Willett *et al.*, 1993).

Following finite convergence, the system rapidly becomes asymmetrical, with one of the two shear zones, the one that developed on the side of the collision to which the velocity discontinuity is attached, accumulating finite strain, with the other only transiently active and accumulating small strain (Willett *et al.*, 1993). The static side of the orogen, i.e. the one to which the discontinuity is attached, is called the retro-side of the orogen; the other, the pro-side (Willett *et al.*, 1993) (Figure 13.2(a)). As deformation proceeds, surface topography is generated, which, for a given set of climatic conditions, can be more or less efficiently eroded away. This results in the net movement of rock particles towards the surface. If the conditions are such that erosion is efficient, a steady-state situation can be

13.1 A simple model for continental collision

(a) Uniform Erosion/Precipitation

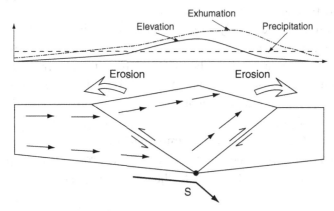

(b) Orographic Precipitation Focussed on Retro-Side

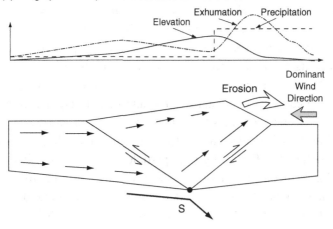

(c) Orographic Precipitation Focussed on Pro-Side

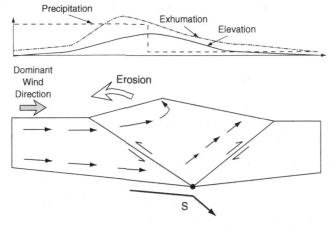

Fig. 13.3. Rock paths in an orogen where erosion is (a) uniform, (b) concentrated on the retro-side of the orogen and (c) concentrated on the pro-side of the orogen.

achieved, in which the collisional flux of rock mass into the orogen is equal to the erosional flux out of the orogen. In this situation the path of rock particles through the orogen is relatively simple, as shown in Figure 13.2.

Rocks enter the orogen at a given depth and first experience a relatively slow increase in pressure and relatively uniform thickening corresponding to their entry into the broad orogenic wedge (Little, 2004); they traverse the pro-shear, where they experience a short but intense pulse of reverse shearing; they then enter the core of the orogen from which they are progressively exhumed by thrusting along the retro-shear. The important point to notice here is that the depth reached by rocks as they 'travel' through the orogen is maximum for those rocks that will be exhumed closest to the surface expression of the retro-shear (Figure 13.2(b)).

The dynamics of this system and, consequently, the geometry of the rock-particle paths are strongly dependent on the nature and efficiency of the erosional processes at the surface of the orogen (Willett *et al.*, 1993; Willett, 1999a), as illustrated in Figure 13.3. For example, if the topography created by the crustal thickening causes a perturbation in the local climatic conditions such that orographic effects enhance the level of precipitation on one of the two sides of the orogen, one can expect greater erosion and thus a higher exhumation rate on the 'wet' side of the mountain belt. This, in turn, can cause the internal force balance within the orogen to shift and lead to important changes in its dynamics and thus in the paths followed by rock particles, as shown in Figure 13.3.

This feedback mechanism has been demonstrated in numerous studies based on numerical models of the coupled tectonics–erosion system (Willett *et al.*, 1993, 2001; Beaumont *et al.*, 1996; Batt and Braun, 1997; Willett, 1999a) and is now widely called upon to explain variations in the surface geology of diverse small mountain belts such as the Southern Alps of New Zealand, Taiwan and the Olympic Mountains.

13.2 Heat advection in mountain belts

Where exhumation of rocks towards the surface advects heat more rapidly than it can be dispersed by conductive heat transport, there is a net transfer of heat to the upper crust, perturbing the thermal structure of the affected region. A one-dimensional consideration of this phenomenon discussed in Chapter 5 showed that efficient heat transport occurs in this way when the rate of exhumation, $\dot{E}$, and the thickness of the layer being exhumed, L, are such that the dimensionless Péclet number $Pe = \dot{E}L/\kappa$ is greater than 1.

Quantification of this process for more complex velocity distributions is not trivial and cannot realistically be achieved using simple analytical solutions such as those developed in Section 5.1. A more pragmatic and flexible approach capable of incorporating the effects of arbitrary material paths and spatial variability in material properties is required, which invariably involves the numerical solution of the heat-transfer equation, as explained in Section 5.3 and Chapter 7. In this section we will simply describe the results obtained in a growing number of studies (Koons, 1987; Shi et al., 1996; Jamieson et al., 1996; Batt and Braun, 1997, 1999) in which the temperature histories of rock particles as they travel through an active mountain belt have been simulated by solving the heat-transport equation numerically, typically in two dimensions and using a velocity field similar to the one shown in Figure 13.2. Such a velocity field can be either prescribed or derived from the numerical solution of the force-balance equation under conditions of continental convergence and surface erosion. Here we will focus on the results of these numerical solutions, not the details of how they were obtained. For details on how one can define a kinematic representation of a tectonic velocity field or calculate it from a dynamic model (by solution of the force-balance equation), the reader is referred to Batt and Braun (1997), Braun and Sambridge (1994) or Herman et al. (2006), among many others.

Advection of heat by the velocity field described in Figure 13.2 leads to an upward deflection of the isotherms in the central region of the orogen, i.e. in the region of maximum exhumation between the two conjugate shear zones (Figure 13.4(b)). The temperature perturbation is partly transferred laterally to the adjacent regions, causing a noticeable heating of the regions on either side of the exhuming orogenic plug and a corresponding reduction of the temperature within the orogenic plug itself. A rock particle passing through the perturbed temperature field first experiences a moderate increase in temperature as it enters the orogen; this increase in temperature is accompanied by an increase in pressure that corresponds to the entry of the particle into the region of thickened crust. Once the particle has made its way through the pro-shear, it experiences rapid decompression during its exhumation; during the early stages of its ascent, however, the particle experiences little cooling because it traverses a nearly isothermal region at the core of the orogen. It is only when it approaches the cold free surface of the orogen that it begins to experience substantial cooling.

This complex history can be illustrated if one plots the trajectory of the particle in pressure–temperature space, as done in Figure 13.4(a) for four particles that will end up at different locations along the surface of the orogen. All particles follow a counter-clockwise P–T trajectory that is made of three main sections: firstly a period of modest increases in temperature and pressure, followed by a period of isothermal decompression and, finally, an episode of rapid cooling with slowly decreasing pressure. The maximum temperature and pressure vary among the particles and increase on going from the pro- to the retro-side of the orogen.

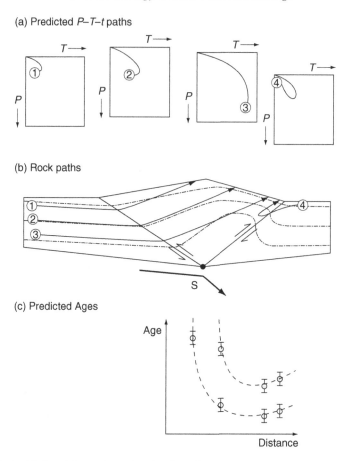

Fig. 13.4. (a) Particle trajectories in (P, T) space as they travel through an orogen as depicted in (b). (c) Predicted age distributions for two arbitrary thermochronological systems.

In the vicinity of the retro-shear, rock particles from either side of the orogen are brought together. This feature of most numerical models explains nicely the formation of inverted metamorphic terrains by tectonic accretion (Jamieson *et al.*, 1996).

The times at which particles exposed at the surface of this model cooled through two arbitrary temperatures are plotted in Figure 13.4(c). This could be viewed as the predicted distribution of cooling ages for a pair of thermochronological systems of different closure temperatures. There is a relatively broad distribution of ages with a minimum in the region of maximum exhumation rate. Ages increase on both sides of this minimum, reaching values that pre-date the onset of the current tectonic event. These ages are those of rock particles that have entered the orogen at a relatively shallow depth and have therefore never experienced sufficiently high temperatures for the particular system to be reset. These particles consequently retain elements of their inherited age structure. Their ages correspond to a previous tectonic event or, simply, to their crystallisation age, which, for the sake of the argument, we will assume is much older than the age of onset of tectonic activity. It follows from these simple geometrical arguments that the zone of reset ages is broader for systems characterised by a lower closure temperature. Furthermore, the higher the closure temperature, the older the ages. These various factors combine to produce an age distribution made of a series of nested 'smiles', one for each thermochronological system.

It is evident from these considerations that the age distribution will be a function not only of the exhumation rate, and thus the horizontal tectonic convergence velocity, but also of the dip of the retro-shear zone and the thickness of the orogen. This demonstrates the potential that thermochronological datasets have to constrain not only the timing of tectonic movements but also the geometry of the deformation patterns they engender (Batt and Braun, 1997). In the following section we will illustrate these points through an example of a rather complete age dataset collected across an active orogen in the South Island of New Zealand.

13.3 The Alpine Fault, South Island, New Zealand

The South Island of New Zealand is the site of an active collision between two continental fragments attached to the Australian and Pacific plates (Figure 13.5). Most of the relative movement between the two plates is constituted by oblique reverse movement along the steeply dipping Alpine Fault (Walcott, 1998).

Figures 13.6 and 13.7 summarise thermochronological data from the study by Batt (1997), which were subsequently published in Batt *et al.* (2000). They include a large range of thermochronometers, ranging from fission tracks in apatite to Ar–Ar in biotite and muscovite. These studies are geographically extensive, comprising a series of transects running perpendicular to the plate boundary and spread laterally over 200 km of the central orogenic region.

Batt and Braun (1997, 1999) developed a fully coupled thermo-mechanical model of the orogenic system to aid in interpreting the dataset and extracting useful information on the development of the tectonic system from it. Using a finite-element approach, the force-balance equation was solved assuming that the crust consists of a complex, non-linear material where brittle failure is active at low pressure and temperature and thermally activated dislocation creep is active at high pressure and temperature. The mechanical model predicted the deformation (and thus velocity) field in the collision zone from a set of imposed velocity boundary conditions that represent subduction of the underlying mantle lithosphere similar to that shown in Figure 13.1. A simple one-dimensional erosion model represents the transport of mass by surface processes. The predicted

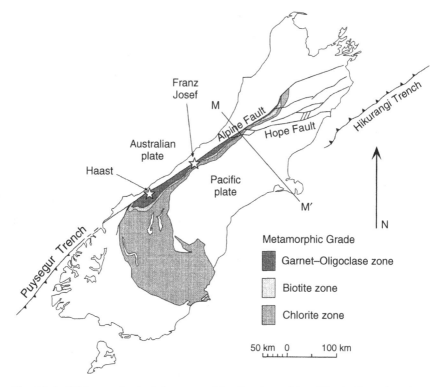

Fig. 13.5. Major geological features of the Southern Alps. The indicated regions of exposed biotite zone and higher metamorphic grade approximately coincide with areas experiencing rapid uplift and exhumation in the ongoing orogeny. The stars labelled Haast and Franz Josef correspond to the locations of published surface heat-flow measurements (Shi *et al.*, 1996) indicating a strong perturbation of the conductive geotherm by advection and, following (5.10), a minimum value for *Pe* of 3.

velocities were used to compute the temperature field by solving the transient heat-conduction/advection/production equation in two dimensions. Particle trajectories were also computed, from which $P-T-t$ paths were calculated. These paths were finally used to compute synthetic ages for a range of thermochronometers. These ages are shown in Figures 13.8 and 13.9.

The predicted ages were shown to display the typical 'nested-smile' pattern described in the previous section: they increase from very young ages adjacent to the surface trace of the retro-shear zone (the equivalent to the Alpine Fault in the Southern Alps) to much older ages within the centre of the orogen. As expected, the minimum age and width of the reset zone vary with the chronometer, as is observed in the data (Figure 13.6).

Exploration of parameter space in the model showed that the pattern of age displayed across the Southern Alps is most consistent with a steeply dipping

13.3 The Alpine Fault, South Island, New Zealand

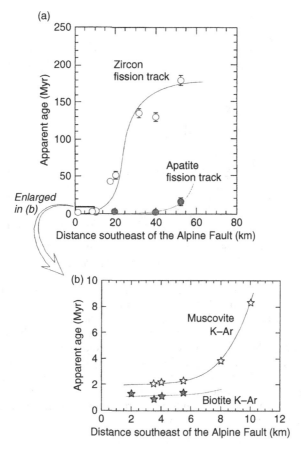

Fig. 13.6. Isotopic age data from Batt (1997). (a) Apatite and zircon fission-track ages from the Whataroa, Perth and Havelock Rivers (all located just north of Franz Josef glacier), plotted against distance southeast of the Alpine Fault. (b) Data plotted on an expanded scale to show variation in muscovite and biotite ages from the same area. Similar patterns of variation across the Southern Alps were reported by Tippett and Kamp (1993) (fission-track ages) and Adams and Gabites (1985) (K–Ar ages). After Batt and Braun (1999). Reproduced with permission from Blackwell Publishing.

(60°) Alpine Fault, the involvement of a crustal layer 25 km thick in the orogenic deformation and convergence rates of $10-12\,\mathrm{mm\,yr^{-1}}$ across the plate boundary.

Futhermore, the increase in age observed along the Alpine Fault from younger ages in the Mount Cook area to older ages towards the southwest (Figure 13.7), is consistent with varied degrees of convergence and thus exhumation along the plate boundary. This variation was, in turn, easily understood by considering the present-day variation in convergence rate predicted from the well-constrained position of the relative pole of rotation between the Pacific and Australian plates in the vicinity of the plate boundary (Batt and Braun, 1999). This led to the

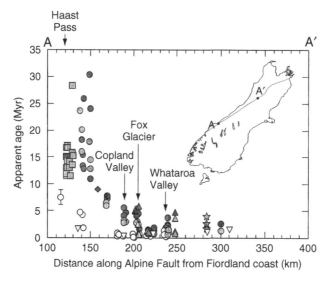

Fig. 13.7. Variation along the Southern Alps in the thermochronological age of material from the zone of consistent young ages close to the Alpine Fault. Muscovite K–Ar ages are indicated by dark grey symbols, biotite K–Ar ages by light grey symbols and zircon fission-track ages by white symbols. Data from Batt (1997) are indicated by circles, data from Adams and Gabites (1985) by squares, data from Adams (1981) by triangles, data from Chamberlain *et al.* (1995) by a diamond, data from Hawkes (1981) by stars and data from Tippett and Kamp (1993) by inverted triangles. Note that the indicated data points for the Tippett and Kamp (1993) study are summaries of a much larger data-set. After Batt and Braun (1999). Reproduced with permission from Blackwell Publishing.

conclusion that the onset of deformation, uplift and exhumation of material at present exposed in the vicinity of the Alpine Fault around Fox Glacier took place approximately 6 Myr ago, a result in very good agreement with independent estimates of the onset of plate convergence (Sutherland, 1996; Walcott, 1998).

13.4 Application of the Neighbourhood Algorithm to Southern Alps data

The results obtained by Batt and Braun (1999) required an extensive search through parameter space to find a set of parameters that would produce a distribution of ages similar to the observations. As seen in Section 8.5, this search can be performed in a more automated and systematic manner by using a so-called inverse method, such as the Neighbourhood Algorithm (NA). Here we will use the dataset of Batt and Braun (1997) described in the previous section to illustrate the use of the NA to constrain the geometry of the Alpine Fault as well as the velocity of convergence between the two plates. We use a simpler model than the

13.4 Application of the NA to Southern Alps data

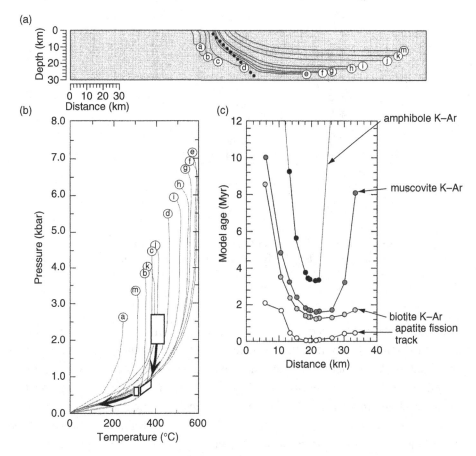

Fig. 13.8. The *P–T* history of material exposed at the surface of the model after 10 Myr. (a) Exhumation trajectories of selected points. The heavy dotted line corresponds to the approximate location of the retro-shear zone in this model. This is inferred to correlate with the Alpine Fault zone for the Southern Alps. (b) *P–T* histories of the points indicated. The *P–T* history of material currently exposed on Craig Peak (Craw *et al.*, 1994) is marked by the bold boxes and black arrows for comparison. (c) Modelled thermochronological ages of the selected points across the surface of the model for a range of different thermochronometers. After Batt and Braun (1999). Reproduced with permission from Blackwell Publishing.

one described above, with a three-dimensional velocity field imposed to represent rock movement along a curved thrust fault and used to compute the temperature field within the crust and the temperature history of particles that travel through the orogen. The computer code used for these calculations is a modified version of the Pecube code (see Chapter 7) in which the three components of the advection term have been included, allowing full representation of a three-dimensional velocity field.

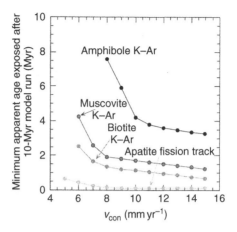

Fig. 13.9. Minimum modelled ages of various isotopic chronometers exposed at the surface of the model after 10 Myr plotted against the velocity of convergence in the latter half of the experiment. After Batt and Braun (1999). Reproduced with permission from Blackwell Publishing.

We first show the results of a single forward model in which the Alpine Fault is assumed to be a listric fault with a surface dip of approximately 60°, soling out at approximately 25 km depth; the rate of convergence is assumed to have abruptly changed from 0.5 to 10 km Myr^{-1} 5 Myr ago and to have remained constant ever since (Batt and Braun, 1999); the mean, conductive surface heat flow is 67 mW m^2. We have used a 1-km-resolution DEM to constrain the geometry of the surface landform. Assuming that rock trajectories are everywhere parallel to the Alpine Fault leads to the surface velocity distribution shown in Figure 13.10(a). The final temperature distribution is shown in Figure 13.10(c). The predicted ages for a set of thermochronometric systems are compared with the data collected along the Whataroa (Batt *et al.*, 2000) and Godley (Tippett and Kamp, 1993) rivers in Figure 13.10(b). The ability of the model to produce the core features of spatial variation in the dataset demonstrates that the tectonic evolution of the area is relatively simple and consistent with the behaviour expected for a simple orogenic system.

The surface dip of the Alpine Fault (here, a simple function of the listricity of the curved fault) and the rate of convergence are also investigated by systematically searching the parameter space by the NA to provide the best fit to the thermochronological data. The listricity, λ, is a length scale that determines the curvature and thus surface dip of the fault according to the following expression for the geometry of the fault:

$$z = f(x) = z_0(1 - e^{-x/\lambda}) \tag{13.1}$$

13.4 Application of the NA to Southern Alps data 205

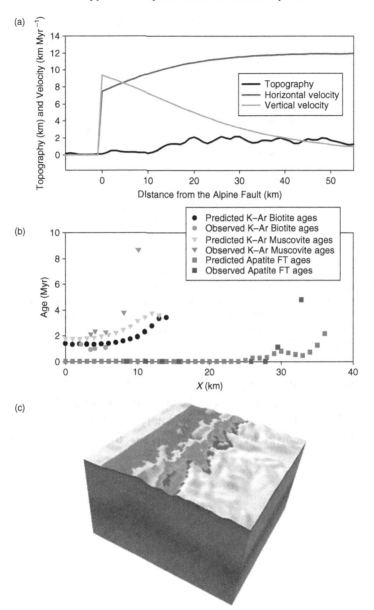

Fig. 13.10. (a) Velocity distributions assumed in order to compute the thermochronological ages shown in (b), where they are compared with real thermochronological data (FT, fission track). (c) The computed three-dimensional temperature structure and surface topography used in the calculations.

where x is the horizontal distance away from the surface trace of the Alpine Fault and z_0 is the soling depth of the fault.

The listricity was allowed to vary between 25 and 50 km, whereas the convergence velocity was bounded between 0 and 15 km Myr^{-1}. The solution of the NA

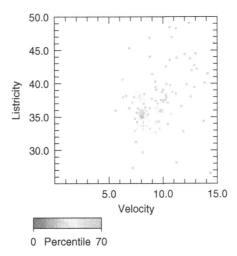

Fig. 13.11. Results of a NA search expressed as locations in the parameter space of the various models tested and their fits. Each dot represents a forward model run; its position corresponds to the values of the parameters used and its colour is proportional to the misfit between predicted thermochronological ages and observations. The cross represents the location of the model run characterised by the smallest misfit.

search is shown in Figure 13.11 and demonstrates that the thermochronological dataset contains information that can be used to constrain the values of both parameters. The best-fit values are approximately $8\,\mathrm{km}\,\mathrm{Myr}^{-1}$ for the convergence rate and 35 km for the listricity (which corresponds to a surface dip of 54° for the Alpine Fault).

Appendix 1

Forward models of fission-track annealing

A1.1 Variable temperature history

Equations (3.15)–(3.18) define isothermal annealing models; they can be applied to a variable temperature history by approximating this as a series of isothermal steps. The amount of annealing at the end of each step is calculated using the 'equivalent-time' algorithm (Duddy et al., 1988); for the fanning Arrhenius model (Equation (3.16)):

$$g(r_j; a, b) = C_0 + \frac{C_1 \ln(t_{eq} + \Delta t_j) + C_2}{1/T_j - C_3} \quad (A1.1.1)$$

where r_j is the reduced track length at the end of the jth isothermal step, Δt_j is the duration of the jth step and T_j is the temperature during the jth step. t_{eq} is the equivalent time, which is defined by

$$\ln(t_{eq}) = -\frac{C_2}{C_1} + \frac{g(r_{j-1}; a, b) - C_0}{c_1}\left(\frac{1}{T_j} - 1\right) \quad (A1.1.2)$$

Equations (3.16) and (A1.1.1) are undefined over certain t–T intervals near $r=0$ and $r=1$ because of the exponents in (A1.1.2). The regions where this occurs are where

$$\ln(t) \leq \frac{-aC_0 - 1}{aC_1}\left(\frac{1}{T} - C_3\right) - \frac{C_2}{C_1} \quad (A1.1.3)$$

near $r=1$ and

$$\ln(t) \geq \frac{(1/b)^a - aC_0 - 1}{aC_1}\left(\frac{1}{T} - C_3\right) - \frac{C_2}{C_1} \quad (A1.1.4)$$

near $r=0$. This problem is circumvented by choosing a time step sufficiently large to fall outside the range (A1.1.3) and stopping the calculations when $t_{eq} + \Delta t$ satisfies (A1.1.4). From there on, $r=0$ is adopted.

The principle of equivalent time presumes that, at any moment, a track that has been annealed to a certain degree behaves during further annealing independently of the conditions which caused the previous annealing. Further annealing depends only on the degree of annealing that has already occurred and the prevailing $t-T$ conditions. The order in which the isothermal steps occur is, therefore, not important.

The isothermal steps are chosen to be of equal length and the associated temperature is obtained by linear interpolation between input points of the thermal history. The annealing history is traced for a set of representative tracks that are formed at the beginning of each new time step. This procedure assumes that the fission-track-production rate is constant through time. The half-life of ^{238}U is sufficiently long (4.45×10^9 yr; Fleischer et al. (1975)) to warrant this assumption, for timescales less than about 500 only Myr. All newly formed tracks share a similar thermal history with existing tracks from their formation onwards. Therefore, the principle of equivalent time allows all of the reduced track lengths to be calculated in a single pass through the isothermal steps in reverse order. In this manner the evolution of the track-length reduction cannot be traced but computing time is greatly reduced (e.g., by a factor of 595 for 100 time steps).

The calculated set of n reduced track lengths is transformed to a binned fission-track-length distribution (FTLD) using a probability-density function $f(r)$. This procedure is required because a given amount of annealing does not lead to a single value of r but to a range in r, as a result of the anisotropy of annealing as well as the range in chemical composition and annealing properties of apatite in the sample (Donelick et al., 1999; Green et al., 1986). Lutz and Omar (1991) proposed the function

$$f(r) = \frac{1}{n} \sum_{j=1}^{n} K\left(\frac{r - r_j}{\sigma}\right) \bigg/ \sigma \qquad \text{(A1.1.5)}$$

where σ is the standard deviation and $K(x)$ is a distribution function. A Gaussian distribution is most applicable:

$$K(x) = \sqrt{\frac{1}{2\pi}} e^{x^2/2} \qquad \text{(A1.1.6)}$$

The standard deviation σ in (A1.1.5) is itself a function of r (Green et al., 1986). Different forms for this function have been suggested. Lutz and Omar (1991) used a three-tier linear relationship with parameter values obtained from a linear

A1.1 Variable temperature history

regression of the data of Green et al. (1986). These values are, however, not well constrained, with correlation coefficients well below 0.5:

$$\sigma = \begin{cases} 2.53 & \text{for } r \leq 0.43 \\ 5.08 - 5.93r & \text{for } 0.43 < r \leq 0.67 \\ 1.39 - 0.61r & \text{for } 0.67 < r \end{cases} \quad (A1.1.7)$$

A histogram of the FTLD can now be constructed by integration of (A1.1.5) over the range of r spanned by each histogram cell, and normalising with respect to 100%.

Calculation of the fission-track age (FTA) is based on the observation of Green (1988) that the contribution to the observed fission-track density, and therefore FTA, of a particular track is dependent on the final length of that track. For n representative tracks produced, the contribution of each track to the total track density is calculated. The relationship between the reduced track density ($d = \rho/\rho_0$) and the reduced track length (r) has been estimated from the data of Green (1988) by Lutz and Omar (1991):

$$d = \begin{cases} 0 & \text{for } r \leq 0.35 \\ 2.15r - 0.76 & \text{for } 0.35 < r \leq 0.66 \\ r & \text{for } 0.66 < r \end{cases} \quad (A1.1.8)$$

The FTA is calculated by multiplying the mean reduced density by the modelled time span (t_a):

$$\text{FTA} = \frac{R_n t_a \sum_{j=1}^n d(r_j)}{n} \quad (A1.1.9)$$

The factor R_n renormalises the calculated FTA to calibrate it against fission-track-age standards, which have a mean length of spontaneous tracks (l_{std}) of about 14.5 μm; $R_n = l_0/l_{\text{std}}$.

Appendix 2

Fortran routines provided with this textbook

Table A2.1. *Stand-alone programs and utility subroutines provided with this textbook*

Program	Functionality
Heat1D	To solve the general transient one-dimensional heat-transport equation
Pecube	To solve the general transient three-dimensional heat-transport equation
Blocking.f	To determine age from a temperature–time history assuming a fixed blocking temperature
Dodson.f90	To determine age from a temperature–time history using Dodson's method
Mad_He.f90	To determine age from a temperature–time history by solving the solid-state diffusion equation by finite difference
MadTrax.f	To determine the fission-track age and track-length distribution from a temperature–time history
Biotite.f90	To determine the K–Ar biotite age from a temperature–time history using Dodson's method
Muscovite.f90	To determine the K–Ar muscovite age from a temperature–time history using Dodson's method
lagtimetoexhum.f	Calculates exhumation rates from thermochronological ages using the approach described in Brandon *et al.* (1998).

Appendix 3

One-dimensional conductive equilibrium with heat production

A3.1 The problem

To understand the effect of heat-producing elements on the conductive heat balance within the continental lithosphere, we assume that a uniform heat production, H, is limited to a crustal layer of thickness δ, the upper limit of which is at a depth h. We assume that asthenospheric convection provides a constant heat flux to the base of the lithosphere. We are going to calculate the temperature at the base of the crust, L, as a function of H, h and δ.

This is an important problem to solve since it is going to tell us how the temperature distribution within the crust is influenced by the heat produced by radioactive elements such as K, Th and U, which are known to be relatively abundant within the upper crust. An interesting application of this problem is the effect on the Moho temperature (and lithospheric strength) of the presence of a strongly heat-producing layer at or near the surface.

A3.2 The basic equation

The one-dimensional equation governing the steady-state transport of heat by conduction can be written as

$$\frac{\partial}{\partial z} k \frac{\partial T}{\partial z} + \rho H = 0 \qquad (A3.2.1)$$

where T is the temperature, z is the vertical spatial coordinate, k is the thermal conductivity, ρ is the density and H is the heat production. This is an elliptic, second-order partial differential equation that requires two boundary conditions. We assume that the surface is kept at a constant, zero temperature:

$$T(z=0) = 0 \qquad (A3.2.2)$$

whereas the base is characterised by a fixed heat flux from the underlying asthenosphere:

$$k\frac{\partial T}{\partial z} = q_m \quad \text{(A3.2.3)}$$

A3.3 The dimensionless form

This equation can be expressed in terms of a dimensionless depth, $z' = z/L$, and temperature, $T' = Tk/(q_m L)$. The temperature scale is derived from the basal boundary condition. This leads to

$$\frac{\partial^2 T'}{\partial z'^2} + H' = 0$$

$$T' = 0 \quad \text{at } z' = 0 \quad \text{(A3.3.1)}$$

$$\frac{\partial T'}{\partial z'} = 1 \quad \text{at } z' = 1$$

where $H' = \rho H L / q_m$.

A3.4 Analytical solution

The crustal column ranging between $z' = 0$ and $z' = 1$ must be divided into three sections. In each section the solution takes a different form. Where there is no heat production, the temperature varies linearly with depth; where there is heat production, the temperature varies quadratically with depth:

$$T' = \begin{cases} a_1 z' + b_1 & \text{for } z' \leq h' \\ -\dfrac{H'}{2} z'^2 + a_2 z' + b_2 & \text{for } h' < z' \leq h' + \delta' \\ a_3 z' + b_3 & \text{for } z' > h' + \delta' \end{cases} \quad \text{(A3.4.1)}$$

where $h' = h/L$ and $\delta' = \delta/L$. The surface boundary condition implies that $b_1 = 0$ whereas the basal boundary condition leads to $a_3 = 1$. To find the values of the other parameters, one imposes that both temperature and heat flux are conserved at the two interfaces $z' = h'$ and $z' = h' + \delta'$. This leads to the following expressions for the dimensionless temperature field:

$$T' = \begin{cases} (1 + H'\delta')z' & \text{for } z' \leq h' \\ -\dfrac{H'}{2} z'^2 + [1 + H'(h' + \delta')]z' - \dfrac{H'h'^2}{2} & \text{for } h' < z' \leq h' + \delta' \\ z' + H'\delta'(h' + \delta'/2) & \text{for } z' > h' + \delta' \end{cases} \quad \text{(A3.4.2)}$$

A3.5 Temperature at the base of the crust

The dimensionless temperature at the base of the crust ($z' = 1$) is given by

$$T'(1) = 1 + H'\delta'(h' + \delta'/2) \qquad (A3.5.1)$$

or, defining $\bar{h}'$ as the depth of the centre of the heat-producing layer,

$$T'(1) = 1 + H'\delta'\bar{h}' \qquad (A3.5.2)$$

which shows that the Moho temperature is a linear function of (a) the amount of heat production, (b) the thickness of the heat-producing layer and (c) the depth of the layer. It is thus important to notice that it is not only the vertically integrated heat production $H'\delta'$ that determines the basal temperature but also the vertical distribution of heat sources.

Note, however, that the dimensionless surface heat flux is given by

$$\left.\frac{\partial T'}{\partial z'}\right|_{z'=0} = 1 + H'\delta' \qquad (A3.5.3)$$

It is proportional to the integrated heat production, $H'\delta'$, but independent of its vertical distribution, h'. This clearly demonstrates that one cannot deduce the temperature structure within the lithosphere/crust from the value of the heat flux (or temperature gradient) at the surface. There is no simple relationship between surface heat flow and the Moho temperature, unless one assumes that all heat sources have the same vertical distribution within the crust.

A3.6 The relationship between heat flux and heat production

By integrating the basic partial differential equation (A3.2.1) over the thickness of the crust, one obtains

$$k \int_0^L \frac{\partial^2 T}{\partial z^2} dz + \int_0^L \rho H \, dz = 0$$

$$k \left.\frac{\partial T}{\partial z}\right|_{z=L} - k \left.\frac{\partial T}{\partial z}\right|_{z=0} + \int_0^L \rho H \, dz = 0 \qquad (A3.6.1)$$

$$-q_0 = -q_m + \int_0^L \rho H \, dz$$

This is another, more direct way to show that the surface heat flow q_0 is simply the sum of the mantle heat flow and the vertically integrated crustal heat production, regardless of its vertical distribution.

Appendix 4

One-dimensional conductive equilibrium with anomalous conductivity

A4.1 The problem

Here we wish to investigate the effect of a non-uniform conductivity distribution. This situation is commonly encountered where sedimentary layers are deposited at the surface of the Earth and, potentially, buried to finite depths. How does this affect the crustal temperature structure, especially the Moho temperature and surface heat flux?

A4.2 The basic equation

We assume that the crustal conductivity is uniform, $k(z) = k_0$, except in a layer of thickness δ of which the top boundary is at a depth h, where the conductivity is k_0/ϵ. Sediments are usually characterised by a lower thermal conductivity than that of crystalline basement, thus ϵ is usually greater than 1. Neglecting heat production, the basic partial differential equation and boundary conditions are

$$\frac{\partial}{\partial z} k(z) \frac{\partial T}{\partial z} = 0$$

$$T = 0 \quad \text{at } z = 0 \tag{A4.2.1}$$

$$k_0 \frac{\partial T}{\partial z} = q_m \quad \text{at } z = L$$

where

$$k(z) = \begin{cases} k_0 & \text{for } z \leq h \\ k_0/\epsilon & \text{for } h < z \leq h+\delta \\ k_0 & \text{for } z > h+\delta \end{cases} \tag{A4.2.2}$$

A4.3 The dimensionless form

This equation can be expressed in terms of a dimensionless depth, $z' = z/L$, and temperature, $T' = Tk/(q_m L)$. The temperature scale is derived from the basal boundary condition. This leads to

$$\frac{\partial}{\partial z'} \frac{k(z)}{k_0} \frac{\partial T'}{\partial z'} = 0$$

$$T' = 0 \quad \text{at } z' = 0 \quad \text{(A4.3.1)}$$

$$\frac{\partial T'}{\partial z'} = 1 \quad \text{at } z' = 1$$

A4.4 Analytical solution

Because we neglect heat production, the solution is made up of three linear temperature distributions:

$$T'(z') = \begin{cases} a_1 z' + b_1 & \text{for } z' \leq h' \\ a_2 z' + b_2 & \text{for } h' < z' \leq h' + \delta' \\ a_3 z' + b_3 & \text{for } z' > h' + \delta' \end{cases} \quad \text{(A4.4.1)}$$

where $h' = h/L$ and $\delta' = \delta/L$. The various coefficients can be derived from the boundary conditions and the continuity of temperature and heat flux across the two interfaces defining the anomalous conductivity layer. This leads to the following expressions for the temperature field in each of the three regions:

$$T'(z') = \begin{cases} z' & \text{for } z' \leq h' \\ \epsilon z' + h'(1 - \epsilon) & \text{for } h' < z' \leq h' + \delta' \\ z' + \delta'(\epsilon - 1) & \text{for } z' > h' + \delta' \end{cases} \quad \text{(A4.4.2)}$$

A4.5 Temperature at the base of the crust

The dimensionless basal temperature is given by

$$T'(1) = 1 + \delta'(\epsilon - 1) \quad \text{(A4.5.1)}$$

It is a function of the conductivity anomaly and the thickness of the anomalous layer but not its position. Furthermore, the surface heat flux is unaffected by the presence of the anomalous layer. This demonstrates that the presence of low-conductivity sediments can strongly affect the temperature at the base of the crust. The depth of burial of the sediments, however, does not affect the Moho temperature and thus the strength of the lithosphere.

Appendix 5

One-dimensional transient conductive heat transport

A5.1 The problem

To illustrate the transient behaviour of conductive systems, we solve the problem of the temporal evolution of the temperature within a crustal layer of thickness L subjected to an instantaneous increase in basal temperature (from T_0 to T_1). This may represent the rapid change in basal crustal temperature accompanying mantle delamination in regions of highly thickened lithosphere such as the Tibetan Plateau. It has been shown that, when the lithosphere is thickened, the lower part of the lithosphere becomes gravitationally unstable and can be rapidly removed by convective asthenospheric flow (Houseman et al., 1981). Here we will assume that this delamination is instantaneous and leads to a step increase in basal temperature.

A5.2 The basic equation

Neglecting heat production by radioactive elements, the one-dimensional equation governing the transient transport of heat by conduction can be written as

$$\rho c \frac{\partial T}{\partial t} = \frac{\partial}{\partial z} k \frac{\partial T}{\partial z} \tag{A5.2.1}$$

where T is temperature, t is time, z is the vertical spatial coordinate, k is the thermal conductivity, ρ is the density and c is the specific heat. This is a parabolic, second-order partial differential equation, which requires two boundary conditions,

$$T(t, z = 0) = 0$$
$$T(t, z = L) = T_1 \tag{A5.2.2}$$

and an initial temperature distribution,

$$T(t = 0, z) = T_0 z / L \tag{A5.2.3}$$

where the ratio

$$T_0/T_1 = \alpha \leq 1 \quad (A5.2.4)$$

represents the assumed jump in temperature imposed for times $t > 0$.

Assuming that the conductivity, density and specific heat are spatially uniform, this equation can be simplified to

$$\frac{\partial T}{\partial t} = \kappa \frac{\partial^2 T}{\partial z^2} \quad (A5.2.5)$$

where $\kappa = k/(\rho c)$ is the thermal diffusivity.

A5.3 The dimensionless form

To illustrate the importance of each term in the equation, one usually re-writes it using the dimensionless variables $z' = z/L$ and $T' = T/T_1$:

$$\frac{\partial T'}{\partial t} = \frac{\kappa}{L^2} \frac{\partial^2 T'}{\partial z'^2} \quad (A5.3.1)$$

which leads to the introduction of the conductive timescale, $\tau_c = L^2/\kappa$, and the further reduction of the equation to its fully dimensionless form:

$$\frac{\partial T'}{\partial t'} = \frac{\partial^2 T'}{\partial z'^2} \quad (A5.3.2)$$

where $t' = t/\tau_c$. The boundary and initial conditions become

$$T'(t', z' = 0) = 0$$
$$T'(t', z' = 1) = 1 \quad (A5.3.3)$$
$$T'(0, z') = \alpha z'$$

A5.4 Analytical solution

First, we redefine the temperature to simplify the basal boundary condition. If we define the new function θ as $\theta = T' - z'$, it is trivial to see that it obeys the same partial differential equation as T',

$$\frac{\partial \theta}{\partial t'} = \frac{\partial^2 \theta}{\partial z'^2} \quad (A5.4.1)$$

but now both boundary conditions are homogeneous:

$$\theta(t', z' = 0) = 0$$
$$\theta(t', z' = 1) = 0 \quad (A5.4.2)$$

The following derivation is based on the variable-separation method, a standard method used to solve partial differential equations with constant coefficients. We assume that the solution can be written as the product of two functions: $\theta = R(t')S(z')$, where R is a function of t' only and S is a function of z' only. Equation (A5.4.1) becomes

$$\frac{\partial(RS)}{\partial t'} = \frac{\partial^2(RS)}{\partial z'^2} \tag{A5.4.3}$$

or

$$\frac{1}{R}\frac{\partial R}{\partial t'} = \frac{1}{S}\frac{\partial^2 S}{\partial z'^2} = -k^2 \tag{A5.4.4}$$

where k is a constant (i.e. does not depend on either t' or z'). Solutions for R and S can be easily found:

$$R = R_0 e^{-k^2 t'}$$
$$S = A\cos(kz') + B\sin(kz') \tag{A5.4.5}$$

On imposing the boundary conditions, we find that

$$A = 0$$
$$\sin k = 0 \tag{A5.4.6}$$

and therefore

$$k = k_n = n\pi \quad \text{for } n = 1, \ldots, \infty \tag{A5.4.7}$$

and the solution takes the form

$$\theta = \sum_{n=1}^{\infty} B_n \sin(n\pi z') e^{-n^2\pi^2 t'} \tag{A5.4.8}$$

Imposing the initial condition leads to

$$\theta(t' = 0) = \sum_{n=1}^{\infty} B_n \sin(n\pi z') = \alpha z' \tag{A5.4.9}$$

Multiplying both sides by $\sin(m\pi z')$ and integrating over the interval $[0, 1]$ leads to

$$\int_0^1 \sum_{n=1}^{\infty} B_n \sin(n\pi z')\sin(m\pi z')\mathrm{d}z' = \int_0^1 \alpha z' \sin(m\pi z')\mathrm{d}z' \tag{A5.4.10}$$

A5.4 Analytical solution

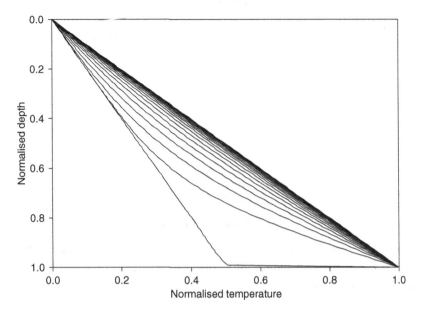

Fig. A5.1. Solution of the equation in dimensional variable space $[z', T']$ for a range of dimensionless time values $t' = 0.025\tau_c k$.

and

$$B_m \int_0^1 \sin^2(m\pi z')\,dz' = \alpha \int_0^1 z' \sin(m\pi z')\,dz'$$

$$B_m \left[\frac{z'}{2} - \frac{\sin(2\pi m z')}{2m\pi}\right]_0^1 = \frac{\alpha}{m\pi}\left[-z\cos(m\pi z') + \frac{\sin(m\pi z')}{m\pi}\right]_0^1 \quad (A5.4.11)$$

$$\frac{B_m}{2} = \frac{\alpha(-1)^m}{m\pi}$$

One can then write

$$\theta = \sum_{n=1}^{\infty} \frac{2\alpha(-1)^n}{n\pi} \sin(n\pi z') e^{-n^2\pi^2 t'} \quad (A5.4.12)$$

or, after re-dimensionalising the variables,

$$T(z,t) = T_1 \frac{z}{L} + \sum_{n=1}^{\infty} \frac{2T_0(-1)^n}{n\pi} \sin\left(\frac{n\pi z}{L}\right) e^{-n^2 t/\tau_c} \quad (A5.4.13)$$

where $\tau_c = \pi^2 L^2/\kappa$ is called the conductive timescale, which is proportional to the square of the thickness of the layer.

The solution is shown graphically in Figure A5.1.

Appendix 6

Volume integrals in spherical coordinates

A6.1 Spherical integrals

The small spherical cube shown in Figure A6.1 has sides given by $[dr, r\cos\theta\, d\phi, r\, d\theta]$. The integral of a function $C(r, \phi, \theta)$ over a sphere of radius R is therefore equal to

$$\int_0^R \int_0^{2\pi} \int_{-\pi/2}^{\pi/2} r^2 \cos\theta\, C(r, \phi, \theta)\, dr\, d\phi\, d\theta \qquad (A6.1.1)$$

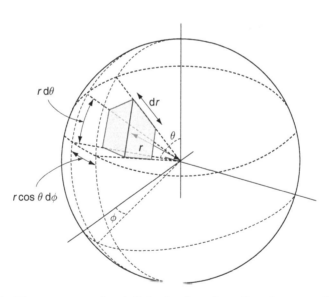

Fig. A6.1. The geometry of an infinitesimal portion of a sphere at a distance r from the centre of the sphere.

A6.1 Spherical integrals

Note that if $C(r, \phi, \theta) = 1$ then this integral reduces to $\frac{4}{3}\pi R^3$, the volume of the sphere. If C is a radial function, i.e. $C = C(r)$, then the integral is reduced to

$$4\pi R^3 \int_0^1 r'^2 C(r') dr' \qquad (A6.1.2)$$

The mean value of the function C over the sphere is the ratio of this integral and the volume of the sphere:

$$\bar{C} = \frac{1}{3} \int_0^1 r'^2 C(r') dr' \qquad (A6.1.3)$$

Appendix 7
The complementary error function

The complementary error function is defined as

$$\operatorname{erfc} x = \frac{2}{\sqrt{\pi}} \int_x^\infty e^{-u^2} \, du \qquad (A7.0.1)$$

for $x \in [0, +\infty]$; it is graphically represented in Figure A7.1.

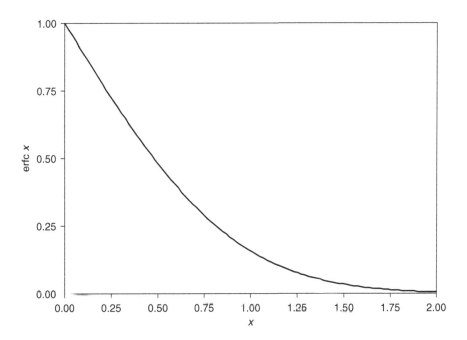

Fig. A7.1. The complementary error function, erfc.

It can be evaluated using the following approximate expressions:

$$\text{erfc } x = 1 - \frac{2}{\sqrt{\pi}}\left(x - \frac{x^3}{3 \times 1!} + \frac{x^5}{5 \times 2!} - \frac{x^7}{7 \times 3!} + \cdots\right)$$
for small values of x

$$\text{erfc } x = \frac{e^{-x^2}}{\sqrt{\pi}x}\left(1 - \frac{1}{2x^2} + \frac{1 \times 3}{(2x^2)^2} - \frac{1 \times 3 \times 5}{(2x^2)^3} + \cdots\right)$$
for large values of x

(A7.0.2)

A more complete description of this function, as well as tabulated values of the function, can be found in Abramowitz and Stegun (1970).

Appendix 8
Pecube user guide

A8.1 How to compile and execute Pecube

There are eight Fortran90 routines in the Pecube package (see Table A8.1). You must have a Fortran90 compiler properly installed on your computer to run Pecube. Pecube comes with a *makefile* file, which can be used on a Unix machine to compile and link the various routines. On other systems, follow the instructions that came with your compiler.

Note that Pecube is a stand-alone package, and does not require any mathematical or graphic library.

Setting up a problem for Pecube can be done in one of two ways: write the information in an input file (*Pecube.in*) created by the user, or use the subroutine *create_pecube_in* to do so. In both cases, Pecube will read the necessary information from an input file called *Pecube.in*. If you wish to use the first option you have to comment out the line *call create_pecube_in* at the top of *Pecube.f90* before compiling, otherwise, Pecube will overwrite the input file you have created!

A8.2 The input file

The input file must contain the following.

1st line: *run*

run is the name of the directory (character*5) where all the output files will be stored; this directory must be created before the run is executed; the input file (*Pecube.in*) will also be copied into this directory.

2nd line: *npe,nsurf, nz, nelemsurf, zl, diffusivity, heatproduction*

(i) *npe* is the number of nodes per two-dimensional element used to discretise the surface topography/landscape (3 = triangular or 4 = rectangular)
(ii) *nsurf* is the number of points defining the landscape (surface *S*)
(iii) *nz* is the number of nodes in the vertical direction (must be an odd number)

A8.2 The input file

Table A8.1. *Routines in the Pecube package*

Routine name	Role	User can edit?
create_pecube_in	Problem definition	Yes
find_temp	P–T–t path calculation	No
interpolate	Vertical interpolation	No
make_matrix	Finite-element matrix formation	No
Peclet_geom	Velocity distribution	Yes
Pecube	Main program	No/Yes
solve_iterative	Gauss–Seidel iterative solver	No
solve_conjugate_gradient	Conjugate-gradient solver	No

(iv) *nelemsurf* is the number of triangles/rectangles connecting the *nsurf* points defining the landscape

(v) *zl* is the thickness of the crust/lithosphere (the depth at which the temperature is fixed at *tmax* (in km))

(vi) *diffusivity* is the thermal diffusivity (in $km^2\,Myr^{-1}$)

(vii) *heatproduction* is the heat-production term (in $°C\,Myr^{-1}$)

3rd line: *tmax, tmsl, tlapse, nstep, ilog, iterative, interpol*

(i) *tmax* is the temperature at the base of the crustal block (at $z = -L$) (in °C)

(ii) *tmsl* is the temperature at mean sea level ($z = 0$) (in °C)

(iii) *tlapse* is the lapse rate (in $°C\,km^{-1}$)

(iv) *nstep* is the number of steps at which you are going to specify the geometry of the surface and the value of the advection velocity

(v) *ilog* is a flag to permit logging ($ilog = 1$) or not ($ilog = 0$) at each time step

(vi) *iterative* is another flag, which determines which solver to use (1 = Gauss–Siedel and 2 = conjugate gradient)

(vii) *interpol* is another flag, which determines which interpolation order (1 or 2) to use during the remeshing step

The logging option should be used only when debugging since it creates an enormous output file (*Pecube.log*) and strongly affects the efficiency of Pecube.

***nsurf* lines:** $(xsurf(i), ysurf(i), i = 1, nsurf)$
xsurf and *ysurf* are real arrays containing the *x*- and *y*-coordinates of the *nsurf* points defining the landscape.

***nelemsurf* lines:** $(iconsurf(k, ie), k = 1, 3/4), ie = 1, nelemsurf)$
iconsurf is the integer connectivity array defining the tessellation to use among the surface nodes. This two-dimensional tessellation forms the basis of the three-dimensional element geometry: the three-dimensional triangular/rectangular prisms are arranged in columns; the top and bottom surfaces of the elements are constructed from the connectivity described by *iconsurf*. The number stored in *iconsurf(k,ie)* is the node number of the *k*th node of the *ie*th triangle, where $k = 1$ to 3/4 and $ie = 1$ to *nelemsurf*. Note that nodes must be stored in a counter-clockwise order for

each element, otherwise the computed element surface is negative and the resulting finite-element matrix is not positive definite.

For each step (and there are *nstep* of them) there should be $1 + nsurf$ lines containing the following.

1st line: *timesurf, v, iout*

  (i) *timesurf* is the time (calculated since the beginning of computations) at which you specify the geometry of the landscape; note that the first step must have $timesurf = 0$
  (ii) *v* is the vertical advection velocity during the period comprised between the current and previous *timesurf* values
  (iii) *iout* is a flag, which determines whether the computed temperature field is saved at this step ($iout = 1$) or not ($iout = 0$)

***nsurf* following lines:** *(zsurf(i), i = 1, nsurf)*
zsurf(i) is the elevation of the surface at location *xsurf(i), ysurf(i)* at time *timesurf*; this is how the geometry of the landscape is defined at every step; Pecube will assume that the surface geometry evolves as a linear function of time between two steps.

A8.3 *create_Pecube_in*

If you choose to use *create_Pecube_in* to generate the input file *Pecube.in*, you can use the template that is provided with Pecube. The template reads in a small part of the topography around Mount Cook derived from a DEM of the South Island of New Zealand (Manaaki Whenua Landcare Research, 1996) and uses it to create a surface topography at a given resolution ($nx \times ny$). You can adjust the values of the various resolution parameters (*nx, ny, nz*) at will, up to $51 \times 51 \times 51$.

You can change the basic parameters such as the diffusivity and the heat production. You can also change the information provided for each step to modify the way you would like the surface topography to evolve with time.

In the template, the surface is held at a constant value (the maximum value read from the DEM) from time 0 to *tcarve*, then the surface landscape is 'carved' into the crustal block from time *tcarve* to *tstop*. The landscape is then allowed to remain unchanged from *tstop* to *tend*.

You can keep the template and simply change the timing of the various events or create your own scenario.

A8.4 *Peclet_geom*

Some users may wish to have the advection velocity vary laterally across the model. To achieve this, a function (*Peclet_geom*) that describes the spatial variation of *v* the advection velocity has been provided. You may modify the function

Peclet_geom at will and re-compile it before execution. The value returned by the function must range between 0 and 1 (otherwise you run the risk of underestimating or overestimating the optimum time step).

Remember that the temporal variation of v is introduced into the model through the *Pecube.in* file (where you specify the geometry of the surface).

A8.5 Output files

If everything goes well and Pecube is run successfully, two output files are created, named *Pecube.out* and *Pecube.ptt*.

Pecube.out is a direct access file which contains two records of length $4*(nsurf*nz)$ bytes per step at which you have requested an output ($iout = 1$) to be saved. The first record contains the z-location of the $nsurf*nz$ nodes; the second record contains the computed temperature of the $nsurf*nz$ nodes.

Note that the nodal information is stored in the following order: $((z((i-1)*nz+k), k = 1, nz), i = 1, nsurf)$ and $((temperature((i-1)*nz+k), k = 1, nz), i = 1, nsurf)$, that is column after column.

Pecube.ptt is a direct-access file that contains *ntime* records, one per time step. Note that this is different from the number of steps (i.e. *nstep*) that you have specified in the input file. The number of time steps cannot be fixed a priori but is a function of the geometry of the problem, the spatial resolution and the values of the various parameters set in the input file. This results from the dynamic determination of the optimum number of time steps to be used in order to preserve accuracy and stability.

Each record in *Pecube.ptt* is $4*(1+2*nsurf)$ bytes long and contains the time, the temperature and the pressure of each rock that will ultimately end up at the surface of the model at time *timesurf(nstep)* (i.e. the last of the times specified in the input file).

Note that the pressure–temperature–time information is stored in the following way: $time, (temperature(i), i = 1, nsurf), (pressure(i), i = 1, nsurf)$.

The last record has *time* set to -1.

Appendix 9
Tutorial solutions

A9.1 Tutorial 1

The computed ages are summarized in Table A9.1. The ages obtained with Mad_He are all very close to 40 Myr, demonstrating that inverting an age dataset to obtain a temperature history is a non-unique problem. Many thermal histories can lead to the same age. Note also that Dodson's method is valid only for simple cooling histories. The absolute-closure-temperature method is almost always inaccurate.

Fission-track-length distributions as computed from MadTrax.f are shown in Figure A9.1. Rapid cooling (scenario 1) leads to a narrow track distribution whereas slow cooling (scenario 2) leads to a broad track distribution.

A9.2 Tutorial 2

Muscovite $^{40}Ar/^{39}Ar$ ages are 40 Myr for the first scenario, and >100 Myr for the four others. Except for the first scenario of very rapid cooling from high temperatures, muscovite $^{40}Ar/^{39}Ar$ is not the appropriate system to study – its closure temperature is too high to discriminate among the different low-temperature thermal histories.

The apatite fission-track ages calculated using MadTrax.f are given in Table A9.1, together with the (U–Th)/He ages calculated in Tutorial 1. The combination of apatite (U–Th)/He and fission-track thermochronometers adds additional constraint to the scenarios and removes some of the ambiguities noted in Tutorial 1. The fission-track ages for the first three scenarios are different.

Fission-track length distributions as computed from MadTrax.f are shown in Figure A9.1. Rapid cooling (scenario 1) leads to a narrow track-length distribution with long mean track length, whereas monotonic slow cooling (scenario 2) leads to a broad and negatively skewed distribution with shorter mean track length.

Table A9.1. *Apparent ages (in Myr) obtained with the various methods for the five thermal scenarios described in Section 2.6: F.D. refers to (U–Th)/He ages determined using Mad_He.f 90, Dodson refers to (U–Th)/He ages determined using Dodson's method with the same diffusion coefficients as used in Mad_He.f 90, Abs. refers to ages obtained by assuming a fixed closure temperature of 75 °C and FT refers to fission-track ages determined using MadTrax.f*

Scenario	F.D.	Dodson	Abs.	FT
1	39.5	41.2	40.1	41.9
2	39.3	43.8	50.0	70.7
3	42.1	22.8	–	93.2
4	40.5	26.4	84.4	91.4
5	39.9	7.6	–	91.9

Note that the final three scenarios with relatively complex histories lead to indistinguishable fission-track ages and length distributions. Resolving these remains ambiguous, even on combining two low-temperature thermochronometers.

A9.3 Tutorial 3

As shown in Appendix 3, the surface heat flow, $-q_0$, is the sum of the contribution from the mantle, $-q_m$, and that from the heat generated within the crust:

$$-q_0 = -q_m + \int_0^L \rho H \, dz \quad (A9.3.1)$$

Assuming that crustal heat production is concentrated in a layer of thickness h_c in the uppermost crust leads to

$$-q_0 = -q_m + h_c \rho H \quad (A9.3.2)$$

showing that there is a linear relationship between heat production and heat flow. The slope of that relationship is the thickness of the heat-producing layer, the intersection (heat flow at zero heat production) is the mantle component to the surface heat flow. A linear regression through the South Australian data (Figure A9.2) yields $-q_m = 28.8 \, \text{mW m}^{-2}$ and $h_c = 9.9 \, \text{km}$. Using Equations (4.36) and (4.37), respectively, we can determine the thermal structure of the crust and the temperature history of the rock. These are shown in Figures 9.2(b) and (c), taking the mean value from Table 4.3 as the heat production ($H = 5.4 \, \mu\text{W m}^{-3}$) and assuming a thermal conductivity $k = 3 \, \text{W m}^{-1}\text{K}^{-1}$, density $\rho = 2.8 \, \text{kg m}^{-3}$ and a surface temperature $T_s = 0 \,°\text{C}$.

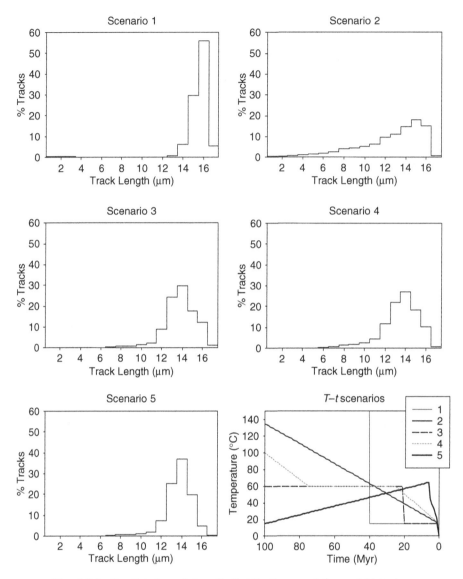

Fig. A9.1. Predicted track-length distributions and thermal histories.

A9.4 Tutorial 4

The very old K–Ar age in muscovite suggests that, during the current tectonic event which started in the Eocene, this rock did not reach temperatures in excess of 400 °C. Because the current tectonic setting is known to have been active since the Eocene, we can also assume that a thermal steady state has been reached. We can therefore use (5.9) to predict age as a function of closure temperature for a range of exhumation rates. We assume that $T_0 = 400\,°\mathrm{C}$ and $L = 15\,\mathrm{km}$. The results are shown in Figure A9.3. An exhumation rate of $1.3\,\mathrm{km\,Myr^{-1}}$ fits the data very well.

A9.4 Tutorial 4

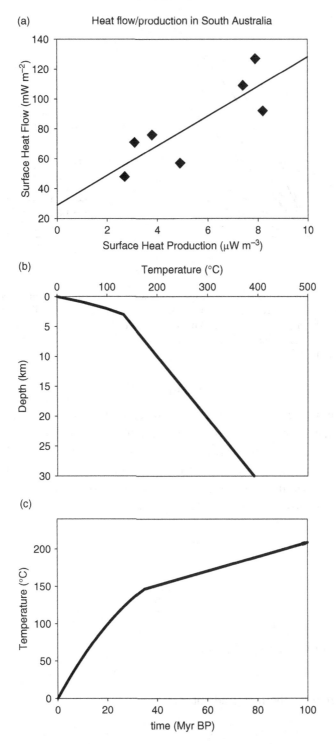

Fig. A9.2. Surface heat flow versus surface-heat-production data from South Australia (diamonds). The line corresponds to a linear regression to the data.

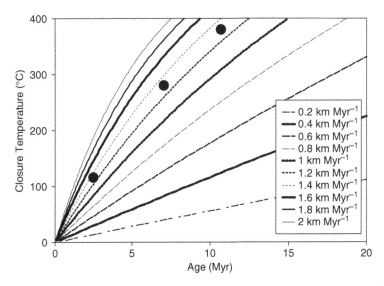

Fig. A9.3. Predicted ages for a range of closure temperatures computed from (5.9). Each curve corresponds to an assumed exhumation rate. The thickness of the layer being exhumed is 15 km and the basal temperature is assumed to be 400 °C. The large black dots correspond to the thermochronological dataset given in Table 5.6.

A9.5 Tutorial 5

The thermochronological data suggest that the current exhumation phase started at least 12 Myr ago. Assuming that this is the time of initiation (t_0), we can postulate that the initial temperature of the rock now at the surface was $T_0 = 375\,°C$. We then compute a series of thermal histories for a range of values of the Péclet number, Pe. These are shown in Figure A9.4. The curve that fits the thermochronological data best corresponds to $Pe = 0.75$. Noting that

$$Pe = \frac{\dot{E}^2 t_0}{\kappa} \quad (A9.5.1)$$

yields

$$\dot{E} = \sqrt{\frac{Pe\,\kappa}{t_0}} = \sqrt{\frac{0.75 \times 25}{12}} = 1.25\,\text{km Myr}^{-1} \quad (A9.5.2)$$

The initial depth of the rock is given by

$$z_0 = \dot{E} t_0 = 1.25 \times 12 = 15\,\text{km} \quad (A9.5.3)$$

and the pre-orogenic thermal gradient was

$$G = \frac{T_0}{z_0} = \frac{375}{15} = 25\,°C\,\text{km}^{-1} \quad (A9.5.4)$$

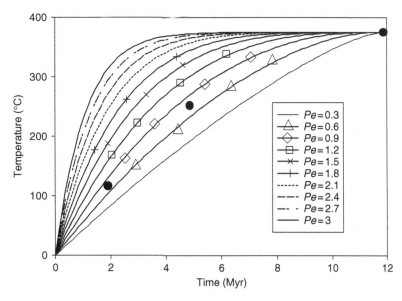

Fig. A9.4. Thermal histories computed from (5.16) assuming a temperature of 375 °C at the time of initiation of exhumation (t_0) of 12 Myr. The circles correspond to the thermochronological data given in Table 5.7.

A9.6 Tutorial 6

Using the package Heat1D, we have computed thermal histories for a rock that ends up at the surface of the Earth in both scenarios. From these thermal histories we can compute the apparent ages for the various systems and, assuming a given closure temperature for each thermochronometer (see Table 1.1), we can construct a temperature–time path. The predicted temperature–time paths for the two scenarios are shown in Figure A9.5. They show that the critical times of 5 and 2 Myr are not represented in these datasets, clearly demonstrating that thermochronology does not provide unequivocal constraints on tectonic events.

A9.7 Tutorial 7

Using Equation (6.11), one can calculate a set of $T-t$ points that can be fed into the subroutine MadTrax.f90 to calculate the apparent fission-track ages of the rock. This operation must be performed for each of the three points (the bottom of the valley, mid-way up the valley wall and the top of the ridge). The apparent exhumation rate can then be estimated by simple linear regression between the three points. This must be done for a series of values of the amplitude and wavelength of the topography. The solution is given in Table A9.2.

The results show that sampling for apatite fission-track dating along a finite-amplitude topography of wavelength smaller than 10 km can provide an accurate

Table A9.2. *Estimates of fission-track ages at* $z = h_0$, $z = 0$ *and* $z = -h_0$, *and apparent exhumation rates*, $\dot{E}_a$, *for various wavelengths*, λ, *and amplitudes*, z_0, *of the surface topography*

	$z_0 = 0.5$ km		$z_0 = 1$ km		$z_0 = 2$ km	
λ (km)	Ages (Myr)	$\dot{E}_a$ (mm yr^{-1})	Ages (Myr)	$\dot{E}_a$ (mm yr^{-1})	Ages (Myr)	$\dot{E}_a$ (mm yr^{-1})
1	5.26 4.81 4.36	1.11	5.71 4.81 3.91	1.11	6.61 4.81 3.04	1.12
10	5.22 4.81 4.37	1.20	5.63 4.81 3.97	1.21	6.43 4.81 3.13	1.21
30	5.07 4.81 4.53	1.86	5.31 4.81 4.24	1.87	5.74 4.81 3.62	1.90

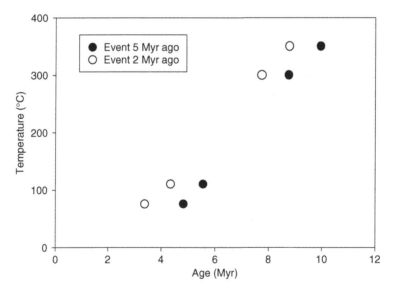

Fig. A9.5. Predicted (U–Th)/He apatite, fission-track apatite and K–Ar biotite and muscovite ages for the two exhumation scenarios, one finishing 5 Myr ago, the other 2 Myr ago.

estimate of exhumation rates in situations where the exhumation rate is of the order of 1 km Myr^{-1}.

The results also demonstrate that the amplitude of the surface topography does not affect the estimate of the exhumation rate but that the range of predicted ages increases with increasing topography. Thus age variations predicted for the small-amplitude topography (0.5 km) might be too small to be measurable in a real situation.

A9.8 Tutorial 8

The computed ages are shown in Figure A9.6. The age–elevation relationship computed from the (U–Th)/He ages is very sensitive to the amount of recent relief change. For cases in which the relief has decreased, the slope of the age–elevation relationship is negative (older ages are found in the bottoms of the

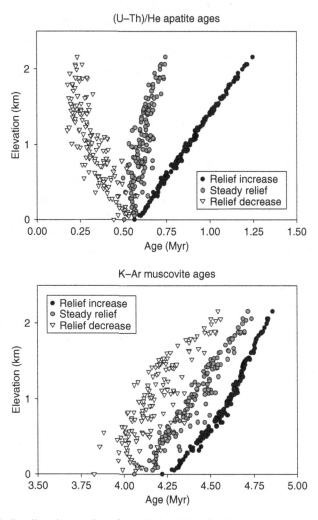

Fig. A9.6. Predicted age–elevation relationships for the three scenarios on relief evolution.

236   *Tutorial solutions*

valleys). The K–Ar muscovite ages are not so sensitive to the change in relief. This is because the closure-temperature isotherm of this system is not affected by the finite-amplitude surface topography.

A9.9 Tutorial 9

(i) According to the parametric trends shown in Figure 10.5, the denudation rate across the Southern Alps ranges from 11 mm yr^{-1} at the Alpine Fault to approximately 1 mm yr^{-1} 30 km to the east. Combined with a lateral convergence rate between the Australian and Pacific Plates of 12 mm yr^{-1}, this gives an η-value of 0.18 for the zircon fission track, 0.36 for the biotite K–Ar and 0.55 for the muscovite K–Ar ages, and 2.44 for the estimate of the 'initial uplift'.

(ii) No direct tectonic significance could be ascribed to the relative age of these systems, since their values indicate that any apparent variation could be influenced either by temporal variation in the orogen, or by the varied time and length scales over which the different systems integrate the laterally variable orogenic denudation rates.

A9.10 Tutorial 10

(i) With admittedly spartan data coverage, no offset can be seen between the erosion-rate distribution across the Olympic Mountains (Figure 10.6) and the variation in apatite (U–Th)/He age across the orogen. For the higher-temperature apatite fission-track system, the minimum in age is offset some 10 km eastward of the peak exhumation rate, whereas for zircon fission-track ages the offset climbs to 30 km. This relative behaviour is consistent with the eastward-directed convergence of the Juan de Fuca Plate and the comparative variability in length scale of constraint and integration indicated by the η-values, with sequentially higher closure temperature producing a corresponding increase in the offset between the local denudation rate and its impact on integrated age.

(ii) If there were no lateral motion of material, λ_x would be zero, consequently making all η-values similarly collapse to zero, indicating that lateral integration of material properties has no influence on thermochronological age. All systems would thus have a common distribution of relative age variation that would correlate directly to the local denudation rate.

References

Abbott, L. D., E. A. Silver, R. S. Anderson *et al.* (1997). Measurement of tectonic surface uplift rate in a young collisional orogen. *Nature*, **385**:501–507.

Abramowitz, M. and I. A. Stegun (1970). *Handbook of Mathematical Functions with Formulas, Graphs, and Mathematical Tables*, 9th edn, New York, Dover.

Adams, C. J. (1981). Uplift rates and thermal structure in the Alpine fault zone and Alpine schists, Southern Alps, New Zealand. In K. R. McClay and N. J. Price, editors, *Thrust and Nappe Tectonics*, Special Publication 9. London, Geological Society of London, pp. 211–222.

Adams, C. J. and A. F. Cooper (1996). K–Ar age of a lamprophyre dike swarm near Lake Wanaka, west Otago, South Island, New Zealand. *New Zealand Journal of Geology and Geophysics*, **39**:17–23.

Adams, C. J. and J. E. Gabites (1985). Age of metamorphism and uplift in the Alpine Schist Group at Haast Pass, Lake Wanaka and Lake Hawea, South Island, New Zealand. *New Zealand Journal of Geology and Geophysics*, **28**:85–96.

Baldwin, S. L., T. M. Harrison and J. D. FitzGerald (1990). Diffusion of ^{40}Ar in metamorphic hornblende. *Contributions to Mineralogy and Petrology*, **105**:691–703.

Barbarand, J., A. Carter, I. Wood and A. Hurford (2003). Compositional and structural control of fission-track annealing in apatite. *Chemical Geology*, **198**:107–137.

Bathe, K.-J. (1982). *Finite Element Procedures in Engineering Analysis*, 1st edn, Englewood Cliffs, New Jersey, Prentice-Hall.

Batt, G. E. (1997). The crustal dynamics and tectonic evolution of the Southern Alps, New Zealand: insights from new geochronological data and fully-coupled thermo-dynamical finite element modeling. Unpublished Ph.D. thesis, The Australian National University, Canberra, ACT.

Batt, G. E., S. L. Baldwin, M. A. Cottam *et al.* (2004). Cenozoic plate boundary evolution in the South Island of New Zealand: new thermochronological constraints. *Tectonics*, **23**:TC4001, doi:10.1029/2003TC001527.

Batt, G. E. and M. T. Brandon (2002). Lateral thinking: 2-D interpretation of thermochronology in convergent orogenic settings. *Tectonophysics*, **394**:185–201.

Batt, G. E., M. T. Brandon, K. A. Farley and M. Roden-Tice (2001). Tectonic synthesis of the Olympic Mountains segment of the Cascadia wedge, using 2-D thermal and kinematic modeling of isotopic age. *Journal of Geophysical Research*, **106**:26 731–26 746.

Batt, G. E. and J. Braun (1997). On the thermomechanical evolution of compressional orogens. *Geophysical Journal International*, **128**:364–382.

(1999). The tectonic evolution of the Southern Alps, New Zealand: insights from fully thermally coupled dynamical modelling. *Geophysical Journal International*, **136**:403–420.

Batt, G. E., J. Braun, B. P. Kohn and I. McDougall (2000). Thermochronological analysis of the dynamics of the Southern Alps, New Zealand. *Geological Society of America Bulletin*, **112**:250–266.

Batt, G. E., B. P. Kohn, J. Braun, I. McDougall and T. R. Ireland (1999). New insight into the dynamic development of the Southern Alps, New Zealand, from detailed thermochronological investigation of the Mataketake Range pegmatites. In U. Ring, M. T. Brandon, G. S. Lister and S. D. Willett, editors, *Exhumation Processes: Normal Faulting, Ductile Flow and Erosion*, London, Geological Society, pp. 261–282.

Beaumont, C., P. Fullsack and J. Hamilton (1992). Erosional control of active compressional orogens. In K. R. McClay, editor, *Thrust Tectonics*, New York, Chapman and Hall, pp. 1–18.

Beaumont, C., P. J. Kamp, J. Hamilton and P. Fullsack (1996). The continental collision zone, South Island, New Zealand: comparison of geodynamical models and observations. *Journal of Geophysical Research*, **101**:3333–3359.

Beaumont, C., H. Kooi and S. Willett (1999). Coupled tectonic–surface process models with applications to rifted margins and collisional orogens. In M. A. Summerfield, editor, *Geomorphology and Global Tectonics*, New York, John Wiley and Sons Ltd, pp. 29–55.

Beck, M. E. (1991). Case for northward transport of Baja and coastal Southern California: paleomagnetic data, analysis, and alternatives. *Geology*, **19**:506–509.

Belytschko, T., H.-J. Yen and R. Mullen (1979). Mixed methods for time integration. *Computer Methods in Applied Mechanics and Engineering*, **17/18**:259–275.

Bernet, M., M. T. Brandon, J. I. Garver *et al.* (2006). Exhuming the Alps through time: clues from detrital zircon fission-track ages. *American Journal of Science*, in press.

Bernet, M., M. T. Brandon, J. I. Garver and B. Molitor (2004). Fundamentals of detrital zircon fission-track analysis for provenance and exhumation studies with examples from the European Alps. In M. Bernet and C. Spiegel, editors, *Detrital Thermochronology – Provenance Analysis, Exhumation and Landscape Evolution of Mountain Belts*, Boulder, Colorado, Geological Society of America pp. 25–36.

Bernet, M., M. Zattin, J. I. Garver, M. T. Brandon and J. A. Vance (2001). Steady-state exhumation of the European Alps. *Geology*, **29**:35–38.

Bierman, P. R. and M. Caffee (2001). Slow rates of rock surface erosion and sediment production across the Namib desert and escarpment, southern Africa. *American Journal of Science*, **301**:326–358.

Bohannon, R. G., C. W. Naeser, D. L. Schmidt and R. A. Zimmerman (1989). The timing of uplift, volcanism and rifting peripheral to the Red Sea: a case for passive rifting? *Journal of Geophysical Research*, **94**:1683–1701.

Brandon, M. T. (1996). Probability density plot for fission-track grain-age samples. *Nuclear Tracks and Radiation Measurements*, **26**:663–676.

Brandon, M. T. and A. R. Calderwood (1990). High-pressure metamorphism and uplift of the Olympic subduction complex. *Geology*, **18**:1252–1255.

Brandon, M. T., M. Roden-Tice and J. I. Garver (1998). Late Cenozoic exhumation of the Cascadia accretionary wedge in the Olympic Mountains, northwest Washington State. *Geological Society of America Bulletin*, **110**:985–1009.

Brandon, M. T. and J. A. Vance (1992). New statistical methods for analysis of fission-track grain-age distributions with applications to detrital zircon ages from the Olympic subduction complex, western Washington State. *American Journal of Science*, **292**:565–636.

Brandt, S. B. and S. N. Voronovskiy (1967). Dehydration and diffusion of radiogenic argon in micas. *International Geology Reviews*, **9**:1504–1507.

Braun, J. (2002a). Estimating exhumation rate and relief evolution by spectral analysis of age-elevation datasets. *Terra Nova*, **14**:210–214.

(2002b). Quantifying the effect of recent relief changes on age–elevation relationships. *Earth and Planetary Science Letters*, **200**:331–343.

(2003). Pecube: a new finite element code to solve the heat transport equation in three dimensions in the Earth's crust including the effects of a time-varying, finite amplitude surface topography. *Computers and Geosciences*, **29**:787–794.

Braun, J. and X. Robert (2006). Constraints on the rate of post-orogenic erosional decay from thermochronological data: example from the Dabie Shan, China. *Earth Surface Processes and Landforms*, in press.

Braun, J. and M. Sambridge (1994). Dynamical Lagrangian Remeshing (DLR): a new algorithm for solving large strain deformation problems and its application to fault-propagation folding. *Earth and Planetary Science Letters*, **124**:211–220.

(1995). A numerical method for solving partial differential equations on highly irregular evolving grids. *Nature*, **376**:655–660.

Braun, J. and P. van der Beek (2004). Evolution of passive margin escarpments: what can we learn from low-temperature thermochronology? *Journal of Geophysical Research*, **109**:F04009, doi:10.1029/2004JF000147.

Brewer, I. D., D. W. Burbank and K. V. Hodges (2003). Modelling detrital cooling-age populations: insights from two Himalayan catchments. *Basin Research*, **15**:305–320.

Brown, R. W. (1991). Backstacking apatite fission track 'stratigraphy': a method for resolving the erosional and isostatic rebound components of tectonic uplift histories. *Geology*, **19**:74–77.

Brown, R. W., K. Gallagher, A. J. W. Gleadow and M. A. Summerfield (2000). Morphotectonic evolution of the South Atlantic margins of Africa and South America. In M. A. Summerfield, editor, *Geomorphology and Global Tectonics*, Chichester, Wiley, pp. 257–283.

Brown, R. W., D. J. Rust, M. A. Summerfield, A. J. W. Gleadow and M. C. J. de Wit (1990). An Early Cretaceous phase of accelerated erosion on the south-western margin of Africa: evidence from apatite fission track analysis and the offshore sedimentary record. *Nuclear Tracks and Radiation Measurements*, **17**:339–350.

Brown, R. W., M. A. Summerfield and A. J. W. Gleadow (1994). Apatite fission track analysis: its potential for the estimation of denudation rates and implications for models of long-term landscape development. In M. J. Kirby, editor, *Process Models and Theoretical Geomorphology*, New York, John Wiley and Sons Ltd, pp. 23–53.

(2002). Denudational history along a transect across the Drakensberg Escarpment of southern Africa derived from apatite fission track thermochronology. *Journal of Geophysical Research*, **107**:2350, doi:10.1029/2001JB000745.

Burbank, D. W. (1992). Causes of recent Himalayan uplift deduced from deposited patterns in the Ganges basin. *Nature*, **357**:680–683.

Burbank, D. W., A. E. Blythe, J. Putkonen *et al.* (2003). Decoupling of erosion and precipitation in the Himalayas. *Nature*, **426**:652–655.

Carlson, W. D. (1990). Mechanisms and kinetics of apatite fission-track annealing. *American Mineralogist*, **75**:1120–1139.

Carlson, W. D., R. A. Donelick and R. A. Ketcham (1999). Variability of apatite fission-track annealing kinetics: I. Experimental results. *American Mineralogist*, **84**:1213–1223.

Carrapa, B., J. Wijbrans and G. Bertotti (2003). Episodic exhumation in the Western Alps. *Geology*, **31**:601–604.

Carslaw, H. S. and C. J. Jaeger (1959). *Conduction of Heat in Solids*, 3rd edn, Oxford, Clarendon.

Carter, A. and C. S. Bristow (2000). Detrital zircon geochronology: enhancing the quality of sedimentary source information through improved methodology and combined U–Pb and fission-track techniques. *Basin Research*, **12**:47–57.

Carter, A. and K. Gallagher (2004). Characterizing the significance of provenance on the inference of thermal history models from apatite fission-track data – a synthetic data study. In M. Bernet and C. Spiegel, editors, *Detrital Thermochronology – Provenance Analysis, Exhumation and Landscape Evolution of Mountain Belts*, Boulder, Colorado, Geological Society of America, pp. 7–23.

Cederbom, C., H. Sinclair, F. Schlunegger and M. Rahn (2004). Climate-induced rebound and exhumation of the European Alps. *Geology*, **32**:709–712.

Cerveny, P. F., N. D. Naeser, P. K. Zeitler, C. W. Naeser and N. M. Johnson (1988). History of uplift and relief of the Himalaya during the past 18 million years: evidence from fission-track ages of detrital zircons from sandstones of the Siwalik group. In K. L. Kleinspehn and C. Paola, editors, *New Perspectives in Basin Analysis*, New York, Springer-Verlag, pp. 43–61.

Chamberlain, C. P., P. K. Zeitler and A. F. Cooper (1995). Geochronologic constraints on the uplift and metamorphism along the Alpine Fault, South Island, New Zealand. *New Zealand Journal of Geology and Geophysics*, **38**:515–523.

Clauser, C. and E. Huenges (1995). Thermal conductivity of rocks and minerals. In T. J. Ahrens, editor, *AGU Handbook of Physical Constants*, Washington, American Geophysical Union, pp. 105–126.

Cliff, R. A. (1985). Isotopic dating in metamorphic belts. *Journal of the Geological Society of London*, **142**:97–110.

Clowes, R. M., M. T. Brandon, A. G. Green et al. (1987). LITHOPROBE southern Vancouver Island: Cenozoic subduction complex imaged by deep seismic reflections. *Canadian Journal of Earth Sciences*, **24**:3151.

Cockburn, H. A. P., R. W. Brown, M. A. Summerfield and M. Seidl (2000). Quantifying passive margin denudation and landscape development using combined fission-track thermochronology and cosmogenic isotope analysis. *Earth and Planetary Science Letters*, **179**:429–435.

Copeland, P. and T. M. Harrison (1990). Episodic rapid uplift in the Himalaya revealed by $^{40}Ar/^{39}Ar$ analysis of detrital K-feldspar and muscovite, Bengal fan. *Geology*, **18**:354–357.

Corrigan, J. (1993). Apatite fission-track analysis of Oligocene strata in South Texas, U.S.A.: testing annealing models. *Chemical Geology*, **104**:227–249.

Coyle, D. A. and G. A. Wagner (1998). Positioning the titanite fission-track partial annealing zone. *Chemical Geology*, **149**:117–125.

Craw, D., M. S. Rattenbury and R. D. Johnstone (1994). Structures within greenschist facies Alpine Schist, Central Southern Alps. *New Zealand Journal of Geology and Geophysics*, **37**:101–111.

Crowley, K. D., M. Cameron and R. L. Schaefer (1991). Experimental studies of annealing of etched fission tracks in fluorapatite. *Geochimica et Cosmochimica Acta*, **89**:1449–1465.

Dahl, P. S. (1996). The effects of composition on retentivity of argon and oxygen in hornblende and related amphiboles: a field tested empirical model. *Geochimica et Cosmochimica Acta*, **60**:3687–3700.

Dahlen, F. A., J. Suppe and D. Davis (1984). Mechanics of fold-and-thrust belts and accretionary wedges: cohesive Coulomb theory. *Journal of Geophysical Research*, **89**:10 087–10 101.

Davis, E. E. and R. D. Hyndman (1989). Accretion and recent deformation of sediments along the northern Cascadia subduction zone. *Geological Society of America Bulletin*, **101**:1465–1480.

Davis, W. M. (1892). The convex profile of badland divides. *Science*, **20**:245.

de Jong, K., J. R. Wijbrans and G. Féraud (1992). Repeated thermal resetting of phengites in the Mulhacen Complex (Betic Zone, southeastern Spain) shown by $^{40}Ar/^{39}Ar$ step heating and single grain laser probe dating. *Earth and Planetary Science Letters*, **110**:173–191.

Densmore, A. L., M. A. Ellis and R. S. Anderson (1998). Landsliding and the evolution of normal-fault-bounded mountains. *Journal of Geophysical Research*, **103**:15 203–15 219.

Dodson, M. H. (1973). Closure temperature in cooling geochronological and petrological systems. *Contributions to Mineralogy and Petrology*, **40**:259–274.

Donelick, R. A., R. A. Ketcham and W. Carlson (1999). Variability of apatite fission-track annealing kinetics: II. Crystallographic orientation effects. *American Mineralogist*, **84**:1224–1234.

Duddy, I. R., P. F. Green and G. M. Laslett (1988). Thermal annealing of fission tracks in apatite 3. Variable temperature behaviour. *Chemical Geology*, **75**:25–38.

Dumitru, T. A., K. Hill, D. Coyle *et al.* (1991). Fission track thermochronology: application to continental rifting of south-eastern Australia. *Australian Petroleum Exploration Association Journal*, **31**:131–142.

Dunai, T. J., A. Bikker and A. G. C. Meesters (2003). DECOMP: a user-friendly forward modelling program for (U–Th)/He low-temperature thermochronology. *Geophysical Research Abstracts*, **5**:14 076.

Ehlers, T. A., P. A. Armstrong and D. S. Chapman (2001). Normal fault thermal regimes and the interpretation of low-temperature thermochronometers. *Physics of the Earth and Planetary Interiors*, **126**:179–194.

Ehlers, T. A., S. D. Willett, P. A. Armstrong and D. S. Chapman (2003). Exhumation of the Central Wasatch Mountains: 2. Thermo-kinematic models of exhumation, erosion and low-temperature thermochronometer interpretation. *Journal of Geophysical Research*, **108**:2173, doi:10.1029/2001JB001723.

England, P. and P. Molnar (1990). Surface uplift, uplift of rocks and exhumation of rocks. *Geology*, **18**:1173–1177.

Farley, K. A. (2000). Helium diffusion from apatite: general behavior as illustrated by Durango fluorapatite. *Journal of Geophysical Research*, **105**:2903–2914.

(2002). (U–Th)/He dating: techniques, calibrations, and applications. In D. P. Porcelli, C. J., Ballentine and R. Wieler, editors, *Noble Gases in Geochemistry and Cosmochemistry*, Washington, Mineralogical Society of America/Geochemical Society, pp. 819–843.

Farley, K. A., R. Wolf and L. Silver (1996). The effect of long alpha-stopping distances on (U–Th)/He dates. *Geochimica et Cosmochimica Acta*, **60**:4223–4229.

Faure, G. (1986). *Principles of Isotope Geology*, 2nd edn, New York, John Wiley and Sons.

Fitzgerald, J. D. and T. M. Harrison (1993). Argon diffusion domains in K-feldspar I: microstructures in MH10. *Contributions to Mineralogy and Petrology*, **113**:36 780.

Fitzgerald, P. G. and A. J. W. Gleadow (1988). Fission-track geochronology, tectonics and structure of the Transantarctic Mountains in northern Victoria Land, Antarctica. *Computers and Geosciences*, **73**:169–198.

Fitzgerald, P. G., J. A. Muñoz, P. J. Coney and S. L. Baldwin (1999). Asymmetric exhumation across the Pyrenean orogen: implications for the tectonic evolution of a collisional orogen. *Earth and Planetary Science Letters*, **173**:157–170.

Fitzgerald, P. G., R. B. Sorkhabi, T. F. Redfield and E. Stump (1995). Uplift and denudation of the central Alaska Range: a case study in the use of apatite fission-track thermochronology to determine absolute uplift parameters. *Journal of Geophysical Research*, **100**:20 175–20 191.

Fleischer, R. L., P. B. Price and R. M. Walker (1975). *Nuclear Tracks in Solids – Principles and Applications*, Los Angeles, University of California Press.

Fleming, A., M. A. Summerfield, J. O. H. Stone, L. K. Fifield and R. G. Creswell (1999). Denudation rates for the southern Drakensberg escarpment, SE Africa, derived from in-situ-produced cosmogenic ^{36}Cl: initial results. *Journal of the Geological Society of London*, **156**:209–212.

Foland, K. A. (1974). Ar-40 diffusion in homogeneous orthoclase and an interpretation of Ar diffusion in K-feldspar. *Geochimica et Cosmochimica Acta*, **38**:151–166.

Förster, A. and H.-J. Förster (2000). Crustal composition and mantle heat flow: implications from surface heat flow and radiogenic heat production in the Variscan Erzgebirge (Germany). *Journal of Geophysical Research*, **105**:27 917–27 938.

Foster, D. A., T. M. Harrison, P. Copeland and M. T. Heizler (1990). Effects of excess argon within large diffusion domains of K-feldspar age spectra. *Geochimica et Cosmochimica Acta*, **54**:1699–1708.

Fowler, C. M. R. (2005). *The Solid Earth: An Introduction to Global Geodynamics*, 2nd edn, Cambridge, Cambridge University Press.

Gaina, C., D. R. Müller, J.-Y. Royer *et al.* (1998). The tectonic history of the Tasman Sea: a puzzle with 13 pieces. *Journal of Geophysical Research*, **103**:12 413–12 434.

Galbraith, R. F. (1990). The radial plot: graphical assessment of spread in ages. *Nuclear Tracks and Radiation Measurements*, **17**:207–214.

Galbraith, R. F. and P. F. Green (1990). Estimating the component ages in a finite mixture. *Nuclear Tracks and Radiation Measurements*, **17**:197–206.

Galbraith, R. F. and G. M. Laslett (1993). Statistical models for mixed fission track ages. *Nuclear Tracks and Radiation Measurements*, **21**:459–470.

(1997). Statistical modelling of thermal annealing of fission tracks in zircon. *Chemical Geology*, **140**:123–135.

Gallagher, K. (1995). Evolving temperature histories from apatite fission track data. *Earth and Planetary Science Letters*, **136**:421–435.

(1997). Thermal history modelling as an inverse problem. In M. F. Middleton, editor, *2nd Nordic Symposium on Petrophysics: Fractured Reservoirs*, Nordic Petroleum Technology Series, Göteborg, pp. 221–245.

Gallagher, K., R. Brown and C. Johnson (1998). Fission track analysis and its applications to geological problems. *Annual Review of Earth and Planetary Sciences*, **26**:519–572.

Gallagher, K. and R. W. Brown (1997). The onshore record of passive margin evolution. *Journal of the Geological Society of London*, **154**:451–457.

Gallagher, K., C. J. Hawkesworth and M. J. M. Mantovani (1994). The denudation history of the onshore continental margin of SE Brazil inferred from apatite fission track data. *Journal of Geophysical Research*, **99**:18 117–18 145.

Garver, J. I. and M. T. Brandon (1994). Fission-track ages of detrital zircon from Cretaceous strata, southern British Columbia: implications for the Baja BC hypothesis. *Tectonics*, **13**:401–420.

Garver, J. I., M. T. Brandon, M. Roden-Tice and P. J. J. Kamp (1999). Exhumation history of orogenic highlands determined by detrital fission-track thermochronology. In U. Ring, M. T. Brandon, S. D. Willett and G. S. Lister, editors, *Exhumation Processes: Normal Faulting, Ductile Flow and Erosion*, London, Geological Society, pp. 283–304.

Garver, J. I. and P. J. J. Kamp (2002). Integration of zircon color and zircon fission track zonation patterns in orogenic belts: application of the Southern Alps, New Zealand. *Tectonophysics*, **349**:203–219.

Garzanti, E., S. Critelli and R. V. Ingersoll (1996). Paleogeographic and paleotectonic evolution of the Himalayan Range as reflected by detrital modes of Tertiary sandstones and modern sands (Indus transect, India and Pakistan). *Geological Society of America Bulletin*, **108**:631–642.

Gautam, P. and Y. Fujiwaral (2000). Magnetic polarity stratigraphy of Siwalik Group sediments of Karnali River section in western Nepal. *Geophysical Journal International*, **142**:812–824.

Gilchrist, A. R., H. Kooi and C. Beaumont (1994). Post-Gondwana geomorphic evolution of southeastern Africa: implications for the controls on landscape development from observations and numerical experiments. *Journal of Geophysical Research*, **99**:12 221–12 228.

Gilchrist, A. R., and M. A. Summerfield (1990). Differential denudation and flexural isostasy in the formation of rifted-margin upwarps. *Nature*, **346**:739–742.

  (1994). Tectonic models of passive margin evolution and their implications for theories of long-term landscape development. In M. J. Kirkby, editor, *Process Models and Theoretical Geomorphology*, New York, Wiley, pp. 221–246.

Giletti, B. J. (1974). Diffusion related to geochronology. In A. W. Hofmann, B. J. Giletti, H. S. Yoder and R. A. Yund, editors, *Geochemical Transport and Kinetics*, Washington, Carnegie Institute, pp. 61–76.

Gleadow, A. J. W. and R. W. Brown (2000). Fission-track thermochronology and the long-term denudational response to tectonics. In M. A. Summerfield, editor, *Geomorphology and Global Tectonics*, New York, Wiley, pp. 57–76.

Gleadow, A. J. W. and I. R. Duddy (1981). A natural long-term track annealing experiment for apatite. *Nuclear Tracks and Radiation Measurements*, **5**:169–174.

Gleadow, A. J. W., I. R. Duddy, P. F. Green and J. F. Lovering (1986). Confined fission track lengths in apatite: a diagnostic tool for thermal history analysis. *Contributions to Mineralogy and Petrology*, **94**:405–415.

Gleadow, A. J. W., B. P. Kohn, R. W. Brown, P. B. O'Sullivan and A. Raza (2002). Fission track thermotectonic imaging of the Australian continent. *Tectonophysics*, **349**:5–21.

Green, P. F. (1981). A new look at statistics in fission-track dating. *Nuclear Tracks and Radiation Measurements*, **5**:77–86.

  (1988). The relationship between track shortening and fission track age reduction in apatite: combined influences of inherent instability, annealing anisotropy, length bias and system calibration. *Earth and Planetary Science Letters*, **89**:335–352.

Green, P. F., I. R. Duddy, A. J. W. Gleadow and J. F. Lovering (1989a). Apatite fission-track analysis as a paleotemperature indicator for hydrocarbon exploration. In

N. D. Naeser and T. H. McCulloh, editors, *Thermal History of Sedimentary Basins*, New York, Springer-Verlag, pp. 181–195.

Green, P. F., I. R. Duddy, A. J. W. Gleadow, P. R. Tingate and G. M. Laslett (1986). Thermal annealing of fission tracks in apatite 1. A qualitative description. *Chemical Geology*, **59**:237–253.

Green, P. F., I. R. Duddy and G. M. Laslett (1988). Can fission track annealing in apatite be described by first-order kinetics? *Earth and Planetary Science Letters*, **89**:216–228.

Green, P. F., I. R. Duddy, G. M. Laslett et al. (1989b). Thermal annealing of fission tracks in apatite 4. Quantitative modelling techniques and extension to geological timescales. *Chemical Geology*, **79**:155–182.

Grimmer, J. C., R. Jonckheere, E. Enkelmann et al. (2002). Cretaceous–Cenozoic history of the southern Tan-Lu fault zone: apatite fission-track and structural constraints from the Dabie Shan (eastern China). *Tectonophysics*, **359**:225–253.

Grove, M. and T. M. Harrison (1996). ^{40}Ar diffusion in Fe-rich biotite. *American Mineralogist*, **81**:940–951.

Gunnell, Y. (2000). Apatite fission track thermochronology: an overview of its potential and limitations in geomorphology. *Basin Research*, **12**: 115–132.

Gunnell, Y., K. Gallagher, A. Carter, M. Widdowson and A. J. Hurford (2003). Denudation history of the continental margin of western peninsular India since the early Mesozoic – reconciling apatite fission-track data with geomorphology. *Earth and Planetary Science Letters*, **215**:187–201.

Gurnis, M., J. X. Mitrovica, J. Ritsema and H. -J. van Heijst (2000). Constraining mantle density structure using geological evidence of surface uplift rates: the case of the African Superplume. *Geochemistry, Geophysics, Geosystems*, **1**:doi:1999GC00035.

Hames, W. E. and S. A. Bowring (1994). An empirical evaluation of the argon diffusion geometry in muscovite. *Earth and Planetary Science Letters*, **124**:161–169.

Hanson, G. N. and P. W. Gast (1967). Kinetic studies in contact metamorphic zones. *Geochimica et Cosmochimica Acta*, **31**:1119–1153.

Harrison, T. M. (1981). Diffusion of ^{40}Ar in hornblende. *Contributions to Mineralogy and Petrology*, **78**:324–331.

Harrison, T. M., P. Copeland, S. A. Hall et al. (1993). Isotopic preservation of Himalayan/Tibetan uplift, denudation, and climatic histories in two molasse deposits. *Journal of Geology*, **100**:157–173.

Harrison, T. M., I. Duncan and I. McDougall (1985). Diffusion of ^{40}Ar in biotite: temperature, pressure and compositional effects. *Geochimica et Cosmochimica Acta*, **49**:2461–2468.

Harrison, T. M., M. Grove, O. M. Lovera and E. J. Catlos (1998). A model for the origin of Himalayan anatexis and inverted metamorphism. *Journal of Geophysical Research*, **103**:27 017–27 032.

Hart, S. R. (1964). The petrology and isotopic mineral age relations of a contact zone in the Front Range, Colorado. *Journal of Geology*, **72**:493–525.

Hawkes, M. R. (1981). The effect of bedrock geology on sediment yield in an alpine area of extreme rainfall. Unpublished Ph.D. thesis, Canterbury University, New Zealand.

Herman, F., J. Braun and W. J. Dunlap (2006). Low temperature themochronology to constrain the tectono-morphic evolution of the Southern Alps of New Zealand. *Journal of Geophysical Research*, in press.

Hess, J. C., H. J. Lippolt, A. G. Gurbanov and I. Michalski (1993). The cooling history of the late Pliocene Eldzhurtinskiy granite (Caucasus, Russia) and the thermochronological potential of grain-size/age relationships. *Earth and Planetary Science Letters*, **117**:393–406.

Hewitt, D. A. and D. R. Wones (1975). Physical properties of some synthetic Fe–Mg–Al trioctahedral biotites. *American Mineralogist*, **60**:854–862.

Hodges, K. V. (1991). Pressure–temperature–time paths. *Annual Review of Earth and Planetary Sciences*, **19**:207–236.

Hodges, K. V., R. R. Parrish and M. P. Searle (1996). Tectonic evolution of the central Annapurna Range, Nepalese Himalayas. *Tectonics*, **15**:1264–1291.

Holmes, A. (1913). *The Age of the Earth*, New York, Harper & Brothers.

House, M. A., K. A. Farley and B. P. Kohn (1999). An empirical test of helium diffusion in apatite: borehole data from the Otway Basin, Australia. *Earth and Planetary Science Letters*, **170**:463–474.

House, M. A., B. P. Wernicke and K. A. Farley (1998). Dating topography of the Sierra Nevada, California, using apatite (U–Th)/He ages. *Nature*, **396**:66–69.

  (2001). Paleo-geomorphology of the Sierra Nevada, California, from (U–Th)/He ages in apatite. *American Journal of Science*, **301**:77–102.

House, M. A., B. P. Wernicke, K. A. Farley and T. A. Dumitru (1997). Cenozoic thermal evolution of the central Sierra Nevada, California, from (U–Th)/He thermochronometry. *Earth and Planetary Science Letters*, **151**:167–169.

Houseman, G. A., D. P. McKenzie and P. Molnar (1981). Convective instability of a thickened boundary layer and its relevance for the thermal evolution of continental convergent belts. *Journal of Geophysical Research*, **86**:6115–6132.

Huang, J. and D. L. Turcotte (1989). Fractal mapping of digitized images: application to the topography of Arizona and comparisons with synthetic images. *Journal of Geophysical Research*, **94**:7491–7495.

Hughes, T. J. R. and A. Brooks (1982). A theoretical framework for Petrov–Galerkin methods with discontinuous weighting functions: applications to the streamline-upwind procedure. In R. H. Gallagher, J. T. Oden, D. H. Norrie and O. C. Zienkiewicz, editors, *Finite Elements in Fluids*, New York, John Wiley and Sons, pp. 47–65.

Hurford, A. J. (1990). International Union of Geological Sciences Subcommission on Geochronology recommendation for the standardization of fission track dating calibration and data reporting. *Nuclear Tracks and Radiation Measurements*, **17**:233–236.

  (1991). Uplift and cooling pathways derived from fission track analysis and mica dating: a review. *Geologische Rundschau*, **80**:349–368.

Hurford, A. J., F. J. Fitch and A. Clarke (1984). Resolution of the age structure of the detrital zircon populations of two Lower Cretaceous sandstones from the Weald of England by fission track dating. *Geological Magazine*, **121**:269–277.

Hurford, A. J. and P. F. Green (1982). A users' guide to fission track dating calibration. *Earth and Planetary Science Letters*, **59**:343–354.

  (1983). The zeta age calibration of fission-track dating. *Chemical Geology (Isotope Geoscience Section)*, **1**:285–317.

Hurley, P. M., H. Hughes, W. H. Pinson and H. W. Fairbairn (1962). Radiogenic argon and strontium diffusion parameters in biotite at low temperatures obtained from Alpine Fault uplift in New Zealand. *Geochimica et Cosmochimica Acta*, **26**:67–80.

Hurley, P. M., E. S. Larsen and D. Gottfried (1956). Comparison of radiogenic helium and lead in zircon. *Geochimica et Cosmochimica Acta*, **9**:98–102.

Huyghe, P., A. Galy, J.-L. Mugnier and C. France-Lanord (2001). Propagation of the thrust system and erosion in the Lesser Himalaya: geochemical and sedimentological evidence. *Geology*, **29**:1007–1010.

Hyndman, R. D. and K. Wang (1993). Thermal constraints on the zone of major thrust earthquake failure: the Cascadia subduction zone. *Journal of Geophysical Research*, **98**:2039–2060.

Jamieson, R. A. and C. Beaumont (1989). Deformation and metamorphism in convergent orogens: a model for uplift and exhumation of metamorphic terrains. In J. S. Daly, R. A. Cliff and B. W. D. Yardley, editors, *Evolution of Metamorphic Belts*, London, Geological Society, pp. 117–129.

Jamieson, R. A., C. Beaumont, J. Hamilton and P. Fullsack (1996). Tectonic assembly of inverted metamorphic sequences. *Geology*, **24**:839–842.

Jamieson, R. A., C. Beaumont, M. H. Nguyen and B. Lee (2002). Interaction of metamorphism, deformation, and exhumation in large convergent orogens. *Journal of Metamorphic Geology*, **20**:9–24.

Jenkins, G. M. and D. G. Watts (1968). *Spectral Analysis and its Applications*, 1st edn, Oakland, California, Holden-Day.

Kamp, P. and J. Tippett (1993). Dynamics of Pacific plate crust in the South Island (New Zealand) zone of oblique continent–continent convergence. *Journal of Geophysical Research*, **98**:16 105–16 118.

Kamp, P. J. J., P. F. Green and J. M. Tippett (1992). Tectonic architecture of the mountain front–foreland basin transition, South Island, New Zealand, assessed by fission track analysis. *Tectonics*, **11**:98–113.

Kasuya, M. and C. W. Naeser (1988). The effect of α-damage on fission-track annealing in zircon. *Nuclear Tracks and Radiation Measurements*, **14**:477–480.

Ketcham, R. A., R. A. Donelick and W. D. Carlson (1999). Variability of apatite fission-track annealing kinetics: III. Extrapolation to geological time scales. *American Mineralogist*, **84**:1235–1255.

Ketcham, R. A., R. A. Donelick and M. B. Donelick (2000). AFTSolve: a program for multi-kinetic modeling of apatite fission-track data. *Geological Materials Research*, **2**(1). URL http://gmr.minsocam.org/papers/v2/v2n1/v2n1abs.html.

King, L. C. (1955). Pediplanation and isostasy: an example from South Africa. *Quaternary Journal of the Geological Society of London*, **111**:353–359.

(1962). *The Morphology of the Earth*, Edinburgh/London, Oliver & Boyd.

Kohn, B. P., D. X. Belton, R. W. Brown et al. (2003). Comment on: "Experimental evidence for the pressure dependence of fission track annealing in apatite" by A. S. Wendt et al. [*Earth Planet. Sci. Lett.* **201** (2002) 593–607]. *Earth and Planetary Science Letters*, **215**:299–306.

Kohn, B. P. and P. Bishop, eds. (1999). Long-term landscape evolution of the southeastern Australian margin: apatite fission track thermochronology and geomorphology. *Australian Journal of Earth Sciences*, **46**:155–232.

Kooi, H. and C. Beaumont (1994). Escarpment evolution on high-elevation rifted margins: insights derived from a surface processes model that combines diffusion, advection and reaction. *Journal of Geophysical Research*, **99**:12 191–12 209.

(1996). Large-scale geomorphology: classical concepts reconciled and integrated with contemporary ideas via a surface processes model. *Journal of Geophysical Research*, **101**:3361–3386.

Koons, P. O. (1987). Some thermal and mechanical consequences of rapid uplift: an example from the Southern Alps, New Zealand. *Earth and Planetary Science Letters*, **86**:307–319.

(1990). Two-sided orogen: collision and erosion from the sandbox to the Southern Alps, New Zealand. *Geology*, **18**:679–682.

(1994). Three-dimensional critical wedges: tectonics and topography in oblique collisional orogens. *Journal of Geophysical Research*, **99**:12 301–12 315.

Kuhlemann, J., W. Frisch, B. Székely, I. Dunkl and M. Kázmér (2002). Post-collisional sediment budget history of the Alps: tectonic versus climatic control. *International Journal of Earth Sciences*, **91**:818–837.

Lan, C.-Y., T. Lee and C. W. Lee (1990). The Rb–Sr isotopic record in Taiwan gneisses and its tectonic implication. *Tectonophysics*, **183**: 129–143.

Laslett, G. M. and R. F. Galbraith (1996). Statistical modelling of thermal annealing of fission tracks in apatite. *Geochimica et Cosmochimica Acta*, **60**:5117–5131.

Laslett, G. M., P. F. Green, I. R. Duddy and A. J. W. Gleadow (1987). Thermal annealing of fission tracks in apatite 2. A quantitative analysis. *Chemical Geology*, **65**:1–13.

Lewis, T. J., W. H. Bentowski, E. E. Davis et al. (1988). Subduction of the Juan de Fuca plate: thermal consequences. *Journal of Geophysical Research*, **93**:15 207–15 225.

Lister, G. S. and S. L. Baldwin (1996). Modelling the effect of arbitrary P–T–t histories on argon diffusion in minerals using the MacArgon program for the Apple Macintosh. *Tectonophysics*, **253**:83–109.

Little, T. A. (2004). Transpressive ductile flow and oblique ramping of lower crust in a two-sided orogen: insight from quartz grain-shape fabrics near the Alpine Fault, New Zealand. *Tectonics*, **23**:doi:10.1029/2002TC001456.

Lonergan, L. and C. Johnson (1998). Reconstructing orogenic exhumation histories using synorogenic detrital zircons and apatites: an example from the Betic Cordillera, SE Spain. *Basin Research*, **10**:353–364.

Lovera, O. M., M. Grove and T. M. Harrison (2002). Systematic analysis of K-feldspar $^{40}Ar/^{39}Ar$ step heating results II: relevance of laboratory argon diffusion properties to nature. *Geochimica et Cosmochimica Acta*, **66**:1237–1255.

Lovera, O. M., M. Grove, T. M. Harrison and K. I. Mahon (1997). Systematic analysis of K-feldspar $^{40}Ar/^{39}Ar$ step heating results: I. Significance of activation energy determinations. *Geochimica et Cosmochimica Acta*, **61**:3171–3192.

Lovera, O. M., M. T. Heizler and T. M. Harrison (1993). Argon diffusion domains in K-feldspar II: kinetic properties of MH10. *Contributions to Mineralogy and Petrology*, **113**:381–393.

Lovera, O. M., F. M. Richter and T. M. Harrison (1989). $^{40}Ar/^{39}Ar$ thermochronology for slowly cooled samples having a distribution of diffusion domain sizes. *Journal of Geophysical Research*, **94**:17 917–17 936.

(1991). Diffusion domains determined by ^{39}Ar released during step heating. *Journal of Geophysical Research*, **96**:2057–2069.

Lutz, T. M. and G. I. Omar (1991). An inverse method of modeling thermal histories from apatite fission-track data. *Earth and Planetary Science Letters*, **104**:181–195.

Manaaki Whenua Landcare Research (1996). *500m Digital Elevation Model for New Zealand*, Lincoln, New Zealand, Manaaki Whenua Landcare Research, Ltd.

Mancktelow, N. S. and B. Grasemann (1997). Time-dependent effects of heat advection and topography on cooling histories during erosion. *Earth and Planetary Science Letters*, **270**:167–195.

Mason, B. (1962). Herschelite – a valid species? *American Mineralogist*, **47**:985–987.

McDougall, I. and T. M. Harrison (1999). *Geochronology and Thermochronology by the $^{40}Ar/^{39}Ar$ Method*, 2nd edn, New York, Oxford University Press.

Meesters, A. G. C. A. and T. J. Dunai (2002a). Solving the production–diffusion equation for finite diffusion domains of various shapes. Part I. Implications for low-temperature (U–Th)/He thermochronology. *Chemical Geology*, **186**:333–344.

(2002b). Solving the production–diffusion equation for finite diffusion domains of various shapes. Part II. Applications to cases with α-ejection and non-homogeneous distribution of the source. *Chemical Geology*, **186**:347–363.

Métivier, F., Y. Gaudemer, P. Tapponier and M. Klein (1999). Mass accumulation rates in Asia during the Cenozoic. *Geophysical Journal International*, **137**:280–318.

Molnar, P. and P. England (1990). Late Cenozoic uplift of mountain ranges and global climate change: chicken and egg? *Nature*, **346**:29–34.

Montgomery, D. R. (1994). Valley incision and the uplift of mountain peaks. *Journal of Geophysical Research*, **99**:13 913–13 921.

Moore, M. A. and P. C. England (2001). On the influence of denudation rates on cooling ages of minerals. *Earth and Planetary Science Letters*, **185**:265–284.

Moore, M. E., A. J. W. Gleadow and J. F. Lovering (1986). Thermal evolution of rifted continental margins: new evidence from fission tracks in basement apatites from southeastern Australia. *Earth and Planetary Science Letters*, **78**:255–270.

Morris, R. G., H. D. Sinclair and A. J. Yelland (1998). Exhumation of the Pyrenean orogen: implications for sediment discharge. *Basin Research*, **10**:69–86.

Müller, W. (2003). Strengthening the link between geochronology, textures and petrology. *Earth and Planetary Science Letters*, **206**:237–251.

Naeser, C. W. (1979). Thermal history of sedimentary basins: fission-track dating of subsurface rocks. In P. A. Scholle and P. R. Schluger, editors, *Aspects of Diagenesis*, Tulsa, Oklahoma, Society of Economic Paleontologists and Mineralogists, pp. 109–112.

  (1981). The fading of fission tracks in the geologic environment: data from deep drill holes. *Nuclear Tracks and Radiation Measurements*, **5**:248–250.

Neumann, N., M. Sandiford and J. Foden (2000). Regional geochemistry and continental heat flow: implications for the origin of the South Australian heat flow anomaly. *Earth and Planetary Science Letters*, **183**:107–120.

Nunn, J. A. and J. R. Aires (1988). Gravity anomalies and flexure of the lithosphere at the Middle Amazon Basin, Brazil. *Journal of Geophysical Research*, **93**:415–428.

Ollier, C. D., ed. (1985). Morphotectonics of passive continental margins. *Zeitschrift für Geomorphologie*, **54**:1–120.

Omar, G. I., M. S. Steckler, W. R. Buck and B. P. Kohn (1989). Fission-track analysis of basement apatites at the western margin of the Gulf of Suez rift, Egypt: evidence for synchroneity of uplift and subsidence. *Earth and Planetary Science Letters*, **94**:316–328.

O'Sullivan, P. B., D. A. Foster, B. P. Kohn, A. J. W. Gleadow and A. Raza (1995). Constraints on the dynamics of rifting and denudation on the eastern margin of Australia: fission track evidence for two discrete causes of rock cooling. In *Proceedings Pacific Rim Conference*, Auckland, pp. 441–446.

Paul, T. A. and P. G. Fitzgerald (1992). Transmission electron microscopy investigation of fission tracks in fluorapatite. *American Mineralogist*, **77**:336–344.

Pazzaglia, F. J. and M. T. Brandon (2001). A fluvial record of long-term steady-state uplift and erosion across the Cascadia forearc high, western Washington State. *American Journal of Science*, **301**:385–431.

Persano, C. (2003). A combination of apatite fission track and (U–Th)/He thermochronometers to constrain the escarpment evolution in south eastern Australia: a case study of high elevation passive margins. Unpublished Ph.D. thesis, University of Glasgow

Persano, C., F. M. Stuart, P. Bishop and D. N. Barford (2002). Apatite (U–Th)/He age constraints on the development of the Great Escarpment on the southeastern Australian passive margin. *Earth and Planetary Science Letters*, **200**:79–90.

Pik, R., B. Marty, J. Carignan and J. Lavé (2003). Stability of the Upper Nile drainage network (Ethiopia) deduced from (U–Th)/He thermochronometry: implications for

uplift and erosion of the Afar plume dome. *Earth and Planetary Science Letters*, **215**:73–88.
Press, W. H., B. P. Flannery, S. A. Teukolky and W. T. Vetterling (1986). *Numerical Recipes, The Art of Scientific Computing*, 1st edn, Cambridge, Cambridge University Press.
Purdy, J. W. and E. Jäger (1976). K–Ar ages on rock-forming minerals from the Central Alps. *Memoirs of the Institute of Geology and Mineralogy of the University of Padova*, **30**:3–31.
Quidelleur, X., M. Grove, O. M. Lovera et al. (1997). Thermal evolution and strike slip history of the Renbu Zedong Thrust, southeastern Tibet. *Journal of Geophysical Research*, **102**:2659–2679.
Rahl, J. M., P. W. Reiners, I. H. Campbell, S. Nicolescu and C. M. Allen (2003). Combined single-grain (U–Th)/He and U/Pb dating of detrital zircons from the Navajo Sandstone, Utah. *Geology*, **31**:761–764.
Rahn, M. K., M. T. Brandon, G. E. Batt and J. I. Garver (2004). A zero-damage model for fission track annealing in zircon. *American Mineralogist*, **89**:473–484.
Rau, W. W. (1973). *Geology of the Washington Coast between Point Grenville and the Hoh River*, technical report, Washington Department of Natural Resources, Geology and Earth Resources Division Bulletin 66.
Ravenhurst, C. E. and R. A. Donelick (1992). Fission track thermochronology. In M. Zentilli and P. H. Reynolds, editors, *Short Course Handbook on Low Temperature Geochronology*, Wolfville, Nova Scotia, Mineralogical Association of Canada, pp. 21–42.
Raymo, M. W. and W. F. Ruddiman (1992). Tectonic forcing of late Cenozoic climate. *Nature*, **359**:117–122.
Reiners, P. W., T. A. Ehlers, S. G. Mitchell and D. R. Montgomery (2003a). Coupled spatial variations in precipitation and long-term erosion rates across the Washington Cascades. *Nature*, **426**:645–647.
Reiners, P. W. and K. A. Farley (1999). Helium diffusion and (U–Th)/He thermochronometry of titanite. *Geochimica et Cosmochimica Acta*, **63**:3845–3859.
Reiners, P. W., K. A. Farley and H. J. Hickes (2002). He diffusion and (U–Th)/He thermochronometry of zircon: initial results from Fish Canyon Tuff and Gold Butte, Nevada. *Tectonophysics*, **349**:297–308.
Reiners, P. W., T. L. Spell, S. Nicolescu and K. A. Zanetti (2004). Zircon (U–Th)/He thermochronometry: He diffusion and comparisons with $^{40}Ar/^{39}Ar$ dating. *Geochimica et Cosmochimica Acta*, **68**:1857–1887.
Reiners, P. W., Z. Zhou, T. A. Ehlers et al. (2003b). Post-orogenic evolution of the Dabie Shan, eastern China, from (U–Th)/He and fission-track thermochronology. *American Journal of Science*, **303**:489–518.
Reyners, M. and H. Cowan (1993). The transition from subduction to continental collision: crustal structure in the North Canterbury region, New Zealand. *Geophysical Journal International*, **115**:1124–1136.
Richter, F. M., O. M. Lovera, T. M. Harrison and P. Copeland (1991). Tibetan tectonics from $^{40}Ar/^{39}Ar$ analysis of a single K-feldspar sample. *Earth and Planetary Sciences Letters*, **105**:266–278.
Ring, U., M. T. Brandon, S. D. Willett and G. S. Lister (1999). Exhumation processes. In U. Ring, M. T. Brandon, S. D. Willett and G. S. Lister, editors, *Exhumation Processes: Normal Faulting, Ductile Flow and Erosion*, London, Geological Society, pp. 1–27.
Robbins, G. A., (1972). Radiogenic argon diffusion in muscovite under hydrothermal conditions. Unpublished M.Sc. Thesis, Brown University.

Rohrman, M., P. A. M. Andriessen and P. A. van der Beek (1996). The relationship between basin and margin thermal evolution assessed by fission track thermochronology: an application to offshore southern Norway. *Basin Research*, **8**:45–63.

Ruiz, G. M. H., D. Seward and W. Winkler (2004). Detrital thermochronology – a new perspective on hinterland tectonics, an example from the Andean Amazon Basin, Ecuador. *Basin Research*, **16**:413–430.

Rutherford, E. (1907). Some cosmical aspects of radioactivity. *Journal of the Royal Astronomical Society of Canada*, 145–165.

Sambridge, M. (1999a). Geophysical inversion with a neighbourhood algorithm – I. Searching a parameter space. *Geophysical Journal International*, **138**:479–494.

(1999b). Geophysical inversion with a neighbourhood algorithm – II. Appraising the ensemble. *Geophysical Journal International*, **138**:727–746.

Sandiford, M. (1999). Mechanics of basin inversion. *Tectonophysics*, **305**:109–120.

Schlunegger, F. (1999). Controls of surface erosion on the evolution of the Alps: constraints from the stratigraphies of the adjacent foreland basins. *International Journal of Earth Sciences*, **88**:285–304.

Schmid, R., T. Ryberg, L. Ratschbacher et al. (2001). Crustal structure of the eastern Dabie Shan interpreted from deep seismic reflection and shallow tomographic data. *Tectonophysics*, **333**:347–359.

Sengör, A. M. C. and K. Burke (1978). Relative timing of rifting and volcanism on Earth and its tectonic implications. *Geophysical Research Letters*, **5**:419–421.

Shi, Y., R. Allis and F. Davey (1996). Thermal modelling of the Southern Alps, New Zealand. *Pure and Applied Geophysics*, **146**:469–501.

Sloan, S. W. (1989). A Fortran program for profile and wavefront reduction. *International Journal for Numerical Methods in Engineering*, **28**:2651–2679.

Small, E. E. and R. S. Anderson (1998). Pleistocene relief production in Laramide mountain ranges, western United States. *Geology*, **26**:123–126.

Smith, E. G. C., T. Stern and B. O'Brien (1995). A seismic velocity profile across the central South Island, New Zealand, from explosion data. *New Zealand Journal of Geology and Geophysics*, **38**:565–570.

Soddy, F. (1911–1914). *Chemistry of the Radio-elements; Parts I and II*, London, Longmans, Green & Co.

Spiegel, C., W. Siebel, J. Kuhlemann and W. Frisch (2004). Toward a comprehensive provenance analysis: a multi-method approach and its implications for the evolution of the central Alps. In M. Bernet and C. Spiegel, editors, *Detrital Thermochronology – Provenance Analysis, Exhumation and Landscape Evolution of Mountain Belts*, Boulder, Colorado, Geological Society of America, pp 37–50.

Stewart, R. J. (1970). Petrology, metamorphism and structural relations of graywackes in the Western Olympic Peninsula, Washington. Unpublished Ph.D. thesis, Stanford University.

Stewart, R. J. and M. T. Brandon (2004). Detrital-zircon fission-track ages for the "Hoh Formation": implications for late Cenozoic evolution of the Cascadia subduction wedge. *Geological Society of America Bulletin*, **116**:60–75.

Stock, J. D. and D. R. Montgomery (1996). Estimating paleorelief from detrital mineral ages. *Basin Research*, **8**:317–327.

Stockli, D. F., K. A. Farley and T. A. Dumitru (2000). Calibration of the apatite (U–Th)/He thermochronometer on an exhumed fault block, White Mountains, California. *Geology*, **28**:961–1056.

Stockli, D. F., J. K. Linn, J. D. Walker and T. A. Dumitru (2001). Miocene unroofing of the Canyon Range during extension along the Sevier Desert Detachment, west-central Utah. *Tectonics*, **20**:289–307.

Stüwe, K., L. White and R. Brown (1994). The influence of eroding topography on steady-state isotherms. Application to fission track analysis. *Earth and Planetary Science Letters*, **124**:63–74.

Suess, E. (1906). *The Face of the Earth (Vol. 2)*, Vienna, F. Tempsky.

Summerfield, M. A. (2000). Geomorphology and global tectonics: introduction. In M. A. Summerfield, editor, *Geomorphology and Global Tectonics*, Chichester, Wiley, pp. 3–11.

Summerfield, M. A. and R. W. Brown (1998). Geomorphic factors in the interpretation of fission-track data. In P. Van den Haute and F. De Corte, editors, *Advances in Fission-Track Geochronology*, Dordrecht, Kluwer Academic Publishers, pp. 269–284.

Sutherland, R. (1996). Transpressional development of the Australia–Pacific boundary through southern South Island, New Zealand: constraints from Miocene–Pliocene sediments, Waiho-1 borehole, South Westland. *New Zealand Journal of Geology and Geophysics*, **39**:251–264.

Tabor, R. W. (1972). Age of the Olympic Metamorphism, Washington: K–Ar dating of low-grade metamorphic rocks. *Geological Society of America Bulletin*, **83**:1805–1816.

Tabor, R. W. and W. H. Cady (1978). *The Structure of the Olympic Mountains, Washington – Analysis of a Subduction Zone*, technical report, U.S. Geological Survey.

Tagami, T., R. F. Galbraith, R. Yamada and G. M. Laslett (1998). Revised annealing kinetics of fission tracks in zircon and geological implications. In P. Van den Haute and F. De Corte, editors, *Advances in Fission-Track Geochronology*, Dordrecht, Kluwer Academic Publishers, pp. 99–112.

Tagami, T., A. J. Hurford and A. Carter (1995). Natural long-term annealing of the zircon fission track system in Vienna Basin deep borehole samples; constraints upon the partial annealing zone and closure temperature. *Chemical Geology*, **130**:147–157.

Tagami, T. and C. Shimada (1996). Natural long-term annealing of the zircon fission track system around a granitic pluton. *Journal of Geophysical Research*, **101**:8245–8256.

ten Brink, U. S. and T. Stern (1992). Rift-margin uplifts and flexural hinterland basins: comparison between the Transantarctic Mountains and the Great Escarpment of southern Africa. *Journal of Geophysical Research*, **97**:569–585.

Thiede, R. C., B. Bookhagen, R. Arrowsmith, E. R. Sobel and M. R. Strecker (2004). Climatic control on rapid exhumation along the Southern Himalayan Front. *Earth and Planetary Science Letters*, **222**:791–806.

Tippett, J. M. and P. J. J. Kamp (1993). Fission track analysis of the late Cenozoic vertical kinematics of continental Pacific Crust, South Island, New Zealand. *Journal of Geophysical Research*, **98**:16 119–16 148.

Tucker, G. E. and R. L. Slingerland (1994). Erosional dynamics, flexural isostasy, and long-lived escarpments: a numerical modeling study. *Journal of Geophysical Research*, **99**:12 229–12 243.

Turcotte, D. L. (1979). Flexure. *Advances in Geophysics*, **21**:51–86.

Turcotte, D. L. and G. Schubert (1982). *Geodynamics: Applications of Continuum Physics to Geological Problems*, 1st edn, New York, John Wiley and Sons.

van der Beek, P. A. (1995). Tectonic evolution of continental rifts: inferences from numerical modelling and fission track thermochronology, Unpublished Ph.D. thesis, Vrije Universiteit, Amsterdam.

van der Beek, P. A., P. A. M. Andriessen and S. Cloetingh (1995). Morpho-tectonic evolution of rifted continental margins: inferences from a coupled tectonic–surface processes model and fission-track thermochronology. *Tectonics*, **14**:406–421.

van der Beek, P. A. and J. Braun (1998). Numerical modelling of landscape evolution on geological time scales: a parameter analysis and comparison with the southeastern highlands of Australia. *Basin Research*, **10**:49–68.

(1999). Controls on post-mid-Cretaceous landscape evolution in the southeastern highlands of Australia: insights from numerical surface process models. *Journal of Geophysical Research*, **104**:4945–4966.

van der Beek, P. A., J. Braun and K. Lambeck (1999). Post-Palaeozoic uplift history of southeastern Australia revisited: results from a process-based model of landscape evolution. *Australian Journal of Earth Sciences*, **46**:157–172.

van der Beek, P. A., S. Cloetingh and P. A. M. Andriessen (1994). Mechanisms of extensional basin formation and vertical motions at rift flanks: constraints from tectonic modelling and fission-track thermochronology. *Earth and Planetary Sciences Letters*, **121**:417–433.

van der Beek, P. A., A. Pulford and J. Braun (2001). Cenozoic landscape evolution in the Blue Mountains (SE Australia): lithological and tectonic controls on rifted margin morphology. *Journal of Geology*, **109**:35–56.

van der Beek, P. A., M. A. Summerfield, J. Braun, R. W. Brown and A. Fleming (2002). Modeling post-breakup landscape development and denudational history across the southeast African (Drakensberg Escarpment) margin. *Journal of Geophysical Research*, **107**: 2351, doi:10.1029/2001JB000744.

van der Wateren, F. M. and T. J. Dunai (2001). Late Neogene passive margin denudation history – cosmogenic isotope measurements from the central Namib desert. *Global and Planetary Change*, **30**:271–307.

Vedder, W. and R. W. T. Wilkins (1969). Dehydroxylation, rehydroxylation, oxidation and reduction of micas. *American Mineralogist*, **54**:482–509.

Vermeesch, P. (2004). How many grains are needed for a provenance study? *Earth and Planetary Science Letters*, **224**:441–451.

von Eynatten, H. and J. R. Wijbrans (2003). Precise tracing of exhumation and provenance using ^{40}Ar/^{39}Ar geochronology of detrital white mica: the example of the central Alps. In T. McCann and A. Saintot, editors, *Tracing Tectonic Deformation Using the Sedimentary Record*, London, Geological Society, pp. 289–305.

Vrolijk, P., R. A. Donelick, J. Queng and M. Cloos (1992). Testing models of fission track annealing in apatite in a simple thermal setting: site 800, leg 129. In R. L. Larson and Y. Lancelot, editors, *Proceedings of the Ocean Drilling Program, Scientific Results, 129*, College Station, Texas, Ocean Drilling Program, pp. 169–176.

Wagner, G. A. (1979). Correction and interpretation of fission track ages. In E. Jäger and J. C. Hunziker, editors, *Lectures in Isotope Geology*, Berlin, Springer-Verlag, pp. 170–177.

Wagner, G. A. and G. M. Reimer (1972). Fission-track tectonics: the tectonic interpretation of fission track ages. *Earth and Planetary Sciences Letters*, **14**:263–268.

Wagner, G. A., G. M. Reimer and E. Jäger (1977). Cooling ages derived from apatite fission track, mica Rb/Sr and K/Ar dating: the uplift and cooling history of the Central Alps. *Memoirs of the Institute of Geology and Mineralogy of the University of Padova*, **30**:1–27.

Wagner, G. A. and P. Van den Haute (1992). *Fission Track Dating*, Amsterdam, Elsevier.
Walcott, R. I. (1998). Modes of oblique compression: Late Cenozoic tectonics of the South Island of New Zealand. *Reviews of Geology and Geophysics*, **36**:1–26.
Wallace, R. C. (1974). Metamorphism of the Alpine Schist, Mataketake Range, South Westland, New Zealand. *Journal of the Royal Society of New Zealand*, **4**:253–266.
Warnock, A. C., P. K. Zeitler, R. A. Wolf and S. C. Bergman (1997). An evaluation of low-temperature apatite U–Th/He thermochronometry. *Geochimica et Cosmochimica Acta*, **61**:5371–5377.
Wellman, H. W. (1979). An uplift map for the South Island of New Zealand, and a model for uplift of the Southern Alps. In R. I. Walcott and M. M. Cresswell, editors, *The Origin of the Southern Alps*, special issue of *Royal Society of New Zealand Bulletin*, 13–20.
Wendt, A. S., O. Vidal and L. T. Chadderton (2002). Experimental evidence for the pressure dependence of fission track annealing in apatite. *Earth and Planetary Science Letters*, **201**:593–607.
Westcott, M. R. (1966). Loss of argon from biotite in a thermal metamorphism. *Nature*, **210**:83–84.
Whipple, K. X. and G. E. Tucker (1999). Dynamics of the stream-power incision model: implications for height limits of mountain ranges, landscape response timescales and research needs. *Journal of Geophysical Research*, **104**:17 661–17 674.
Willett, S. D. (1997). Inverse modeling of annealing of fission tracks in apatite 1; a constrained random search method. *American Journal of Science*, **297**:939–969.
  (1999a). Orogeny and orography: the effects of erosion on the structure of mountain belts. *Journal of Geophysical Research*, **104**:28 957–28 981.
  (1999b). Rheological dependence of extension in wedge models of convergent orogens. *Tectonophysics*, **305**:419–435.
Willett, S. D., C. Beaumont and P. Fullsack (1993). Mechanical model for the tectonics of doubly-vergent compressional orogens. *Geology*, **21**:371–374.
Willett, S. D. and M. T. Brandon (2002). On steady states in mountain belts. *Geology*, **30**:175–178.
Willett, S. D., D. Fisher, C. Fuller, Y. En-Chao and L. Chia-Yu (2003). Erosion rates and orogenic-wedge kinematics in Taiwan inferred from fission-track thermochronometry. *Geology*, **31**:945–948.
Willett, S. D., D. R. Issler, C. Beaumont, R. A. Donelick and A. M. Grist (1997). Inverse modeling of annealing of fission tracks in apatite 2; application to the thermal history of the Peace River Arch Region, Western Canada Sedimentary Basin. *American Journal of Science*, **297**:970–1011.
Willett, S. D., R. Slingerland and N. Hovius (2001). Uplift, shortening, and steady state topography in active mountain belts. *American Journal of Science*, **301**:455–485.
Wobus, C. W., K. V. Hodges and K. X. Whipple (2003). Has focused denudation sustained active thrusting at the Himalayan topographic front? *Geology*, **31**:861–864.
Wolf, R. A., K. A. Farley and D. M. Kass (1998). Modeling of the temperature sensitivity of the apatite (U–Th)/He thermochronometer. *Computers and Geosciences*, **148**:105–114.
Wolf, R. A., K. A. Farley and L. T. Silver (1996). Helium diffusion and low-temperature thermochronometry of apatite. *Geochimica et Cosmochimica Acta*, **60**:4231–4240.
Yamada, R., T. Tagami, S. Nishimura and H. Ito (1995). Annealing kinetics of fission tracks in zircon: an experimental study. *Chemical Geology*, **122**:249–258.

Zaun, P. E. and G. A. Wagner (1985). Fission track stability in zircons under geological conditions. *Nuclear Tracks and Radiation Measurements*, **10**:303–307.

Zeitler, P. K., A. L. Herczig, I. McDougall and M. Honda (1987). U–Th–He dating of apatite: a potential thermochronometer. *Geochimica et Cosmochimica Acta*, **51**:2865–2868.

Zeitler, P. K., N. M. Johnson, C. W. Naeser and R. A. K. Tahirkheli (1982). Fission-track evidence for Quaternary uplift of the Nanga Parbat region, Pakistan. *Nature*, **298**:255–257.

Zeitler, P. K., A. S. Meltzer, P. O. Koons *et al.* (2001). Erosion, Himalayan geodynamics, and the geomorphology of metamorphism. *GSA Today*, **11**:4–9.

Zhang, P., P. Molnar and W. R. Downs (2001). Increased sedimentation rates and grain sizes 2–4 Myr ago due to the influence of climate change on erosion rates. *Nature*, **410**:891–897.

Zienkiewicz, O. C. (1977). *The Finite Element Method*, 3rd edn, Maidenhead, McGraw-Hill.

Index

activation energy, 141
admittance, 123, 126, 129
advection, 7, 60, 64, 65, 68, 76–79, 81, 116, 117, 119, 196, 197
    lateral, 8, 116
    timescale, 79, 96
age
    apparent, 23, 28, 151
    central, 51, 149
    cooling, 132
    crystallisation, 21
    depositional, 139
    detrital, 140, 144, 145
    fission-track, 51
    peak, 139, 149
    radiometric, 20
    single-grain, 51, 53, 133, 139, 140, 142, 144, 146
    standard, 36, 209
    stratigraphic, 132, 149
age–elevation relationship, 6–9, 110, 111, 113, 120–122, 123, 127, 144, 145, 147, 153, 164, 170–172, 188
algebraic system, 29
α-damage, 54
α-ejection, 17, 44
Alpine Fault, 124, 199, 200, 202, 204
argon release, 36
atmospheric argon, 34

Bega Valley, 189
blanketing effect, 71, 181
blocking temperature, 23
boundary conditions, 30, 66, 67, 105
    Dirichlet, 66, 84, 99
    Neumann, 66, 84, 85

Cascade, 177, 187
Cascadia, 157, 158, 160
chemical etching, 48
Cholesky factorization, 98–100
closure depth, 141

closure temperature, 4, 13, 23–26, 31, 41, 42, 54, 69, 80, 111, 113, 122, 126, 129, 141, 143, 145, 148, 154, 160, 170, 198, 199
closure time, 25
compensation depth, 164
complementary error function, 82
component ages, 139
component populations, 139
conduction, 7, 60, 63, 65, 68, 77–79
    equation, 63
    timescale, 79, 96
conductivity, 63, 66, 69, 70, 78, 82, 105, 140, 181
conjugate shear zones, 197
continental collision, 192
convection, 60
cooling rate, 24, 26
Crout's method, 100

Dabie Shan, 169, 171
damage zone, 48
decay constant, 19, 20
    fission-track, 49
    helium, 43
decay reaction
    Ar–Ar, 35
    K–Ar, 33
    U–Th–He, 43
density, 66, 78, 82
denudation, 9, 12, 151, 152, 180, 182, 186
denudation rate, 8, 131, 133, 134, 139–142, 144, 151, 155, 158, 160, 171
    spatial variability, 155
description
    Eulerian, 64, 65
    Lagrangian, 64, 116
detrital grains, 132
detrital thermochronology, 9, 131
diffusion, 1, 3, 19, 21
    Ar, 39
    biotite, 40
    coefficients, 26
    constant, 21
    domain, 21, 31

diffusion (cont.)
  equation, 21
  feldspar, 39
  helium, 42
  hornblende, 41
  hydrothermally buffered, 40
  muscovite, 41
  parameters, 21, 30, 47
  phlogopite, 40
  solid-state, 1, 20, 27
    numerical solution, 27
  volume, 40
diffusivity, 21, 23, 24, 28, 78, 83, 107, 119, 125, 141
  argon, 39
  helium, 44, 45
divergence, 63
Dodson, 24, 25
doubly-vergent critical wedge model, 194
downwearing
  erosional, 179
Drakensberg Escarpment, 183

elastic thickness, 166
  effective, 164, 169, 171, 172, 174
equation
  age equation, 30
  age–elevation slope, 113
  algebraic system
    tri-diagonal, 30
  annealing, 57–59
  Ar–Ar age, 36
  Arrhenius, 21
  boundary condition, 99, 115
    Dirichlet, 66
    Neumann, 66
  closure depth, 141
  closure temperature, 26
  conduction
    periodic topography, 106
    steady-state, 68, 105
    transient, 63
    variable conductivity, 70
  conduction–advection, 65
    dimensionless, 78
    periodic topography, 108, 110
    steady-state, 77, 106
    transient, 81
  conduction–advection–production, 65
  conduction–production
    steady-state, 73
  denudation rate
    spatial variability, 155
  detrital thermochronology
    distribution, 137
  diffusivity, 21, 78
  dimensionless form, 78, 79, 83
  energy conservation, 63
  error function
    complementary, 82
  η-factor, 155

  exponential production
    steady-state, 75
  F-statistics, 138
  factorization
    Cholesky, 99
    Crout, 100
  fault geometry, 204
  finite-difference operators, 29
  finite element, 87, 96
    assembly, 97
  fission-track age, 49, 50
  flexural rigidity, 12, 166
  Fourier's law, 63
  gain, 124
  general heat transport, 66, 115
  heat, 65, 66, 68, 69, 72, 77, 81, 83, 105
    transient, 83
  integral, 84, 86
  irradiation factor, 36
  isostatic compensation factor, 168
  isostatic rebound, 12, 165
  isotopic age, 20
  Jacobian, 88
  K–Ar age, 34
  misfit, 173
  numerical integration
    Gauss quadrature, 92
    Newton–Cotes, 92
  parallel Arrhenius, 55
  partial differential, 84
  Péclet number, 79
    elemental, 96
  Petrov–Galerkin, 117
  radioactive decay, 19
  relief change, 126
  rock uplift, 12
  shape function, 85, 87
    derivative, 90
    linear, 88
    quadratic, 89
  solid-state diffusion, 21
    finite difference, 29
    spherical, 28
  Taylor expansion, 29
  thin-plate deflection, 166
  time integration
    explicit–implicit, 95
    implicit–explicit, 29
  timescale
    advection, 96
    conduction, 96
  trapezoidal rule, 30
  ζ factor, 50
erosion, 7, 152, 164, 192
erosion cycles, 179
erosion rate, 171
escarpment, 177
escarpment retreat, 178
escarpment-retreat scenario, 184, 187, 191
etching, 55
European Alps, 139, 145
excess argon, 35, 36

Index

exhumation, 1, 7–9, 13, 76, 77, 79–81, 83, 106, 111, 119, 151, 152, 201
   path, 153, 154
exhumation rate, 14, 77, 78, 80–82, 108, 111, 121, 124–127, 129, 132, 140, 141, 144, 145, 147–149, 151, 155, 159, 174, 196, 198, 199
external-detector method, 50, 51

finite difference, 28
   equation, 29
finite element, 84, 116, 119, 199
   discretization, 86
   equation, 84, 96, 98, 117
   matrix, 96
   method, 83, 84
   two-dimensional, 89
fission
   induced, 50
   spontaneous, 20, 49
fission track, 111, 160
   activation energy, 55
   annealing, 52–54
   annealing model, 55
      curvilinear Arrhenius, 55
      fanning Arrhenius, 55
      multi-compositional, 58
      parallel Arrhenius, 55
   length distribution, 53, 54
flexural compensation, 164
flexural isostasy, 12, 164, 168, 179
flexural rigidity, 166, 188, 189
flexural strength, 168
flexural wavelength, 167
fluid inclusion, 18, 35
foreland basin, 132
Fortran programs, 210
   Biotite.f90, 41
   Dodson.f, 32
   Heat1D.f, 102
   lagtimetoexhum.f, 141
   Mad_He.f90, 48
   MadHe.f, 32
   MadTrax.f, 59
   Muscovite.f90, 41
   Pecube.f90, 224
Fourier transform, 124
Fourier's law, 63
French Alps, 147

gain, 124, 126, 127
Galerkin method, 86
gas constant, 21
Gauss quadrature, 92, 93
Gauss–Siedel method, 101
geometrical constants, 26
geomorphology, 43
geothermal gradient, 7, 69, 70, 81, 82, 125, 129, 140, 181, 188, 191
Godley River, 204

half-life, 20, 21
   spontaneous fission, 49

heat capacity, 66, 78, 82
heat flux, 61, 66
   surface, 81, 204
heat production, 65, 66, 72–74, 82, 105, 125, 141
Heat1D, 102
Himalayan collision, 144
Himalayas, 132
hypsometry, 145, 147

inclusions
   in apatite, 44
integral equation, 85, 86
inverse methods, 122, 130
inversion, 173, 189
irradiation, 35
irradiation factor, 36
isostasy
   principle of, 164
isostatic amplification factor, 165
isostatic compensation degree of, 168
isostatic rebound, 11, 12, 164, 166, 167, 182, 183
isotopic systems, 3, 19
iterative method, 100, 101

Jacobi method, 101
Jacobian, 88
Juan de Fuca plate, 157, 158

Kings Canyon, 121, 127

laboratory experiments, 31
lag time, 132, 140, 141
landscape-evolution model, 179, 182, 185, 199
Laplacian, 21, 64
lapse rate, 67, 116
Laramide Orogeny, 118, 119, 127
lateral advection, 151
lateral motion, 156, 158
Legendre polynomials, 93
listricity, 205
lithology, 14
lithosphere, 61
local isostasy, 12

magmatic underplating, 181
metamorphic minerals, 39
method
   Ar–Ar, 4, 5, 16, 31, 33, 35, 36, 39, 42, 132, 136, 148
   fission-track, 4, 5, 8, 15–17, 20, 33, 48–50, 52–54, 57, 59, 132, 136, 137, 148, 165
   K–Ar, 4, 16, 33–36, 41, 199, 201, 202
   U–Th–He, 4, 5, 14, 15, 17, 27, 33, 42–45, 47
mineral pairs, 39
mineralogy, 14
misfit function, 130, 173, 206
Monte Carlo method, 130
Mount Cook, 201
multiple-diffusion domain, 39

Neighbourhood Algorithm, 130, 173, 189, 202, 206
Nepal, 148

neutron fluence, 36
Newton–Cotes formula, 92
nuclear decay, 1
numerical integration, 92

Olympic Mountains, 157, 158, 196
open-system temperature, 22
orogenic phases, 133, 134
orographic effects, 196
over-crushing, 16

paleo-PAZ, 150
paleo-relief, 144
partial-annealing zone, 52, 191
partial resetting, 131, 139, 148
partial retention, 23
partial-retention zone, 8, 23, 24
   helium, 47
passive margin, 177, 179, 182, 186,
Péclet number, 78–81, 83, 96, 196
Pecube, 116–118, 125, 177, 187, 203
periodic surface load, 167
Petrov–Galerkin method, 117
photometry, 34
planation, 179
plateau-degradation scenario, 184, 187, 191
power spectra, 126
probability density
   function, 143, 144, 147, 208
   plot, 142
production, 65
production rate, 21
P–T–t path, 197, 200

radioactive decay, 19, 20
radioactive elements, 72
reduced integration, 93
relief, 113, 131, 144–146, 148, 192
relief reduction, 127
residue, 101
rifting, 181
rock uplift, 7

sample preparation, 16
sampling strategy, 13, 129
sediment provenance, 132
sedimentary basin, 132, 139, 148
sedimentary recycling, 144
sedimentology, 132
selective integration, 94
shape function, 85, 87–89
   derivative, 90
shear
   pro-, 196
   retro-, 196, 197, 199
Sierra Nevada, 118, 127
skin depth, 106
skyline storage, 98
South Island, New Zealand, 124, 199

Southern Alps, New Zealand, 132, 192, 196
spectral method, 122, 124, 129
spot fusion, 36
stationary datasets, 129
steady state
   age, 13
   exhumational, 76, 139, 140
   flux, 134, 194
   heat, 211
   solution, 76
   thermal, 8, 14, 68, 72, 74, 76, 77, 91, 105, 106, 110, 111, 127, 140, 144, 146, 229
   topographic, 14, 145, 146
step heating, 36, 39
stratigraphy, 132
structural control, 154, 199
structures, 14
surface processes, 1, 154, 199
   fluvial incision, 183
   hillslope transport, 183
   landsliding, 183
   stream power law, 183
surface relief, 118, 126, 127, 129
surface topography, 105, 106, 108, 109, 111, 122, 123, 126, 129, 194

Taiwan, 132, 196
Tasman Sea, 189
Taylor expansion, 29
tectonic assembly, 154
tectonics, 1, 192
temperature
   gradient, 63, 77, 106
thermal diffusivity, 78, 82
thermal history, 157
thermochronometer, 21
   choice of, 13
time integration
   explicit method, 29, 94, 95
   implicit method, 29, 94–96
time stepping, 94
track-length distribution, 148
transient, 81, 83, 115
trapezoidal rule, 30
tri-diagonal system, 30

uplift, 151, 164
   isostatic, 165
   rock, 10, 77
   surface, 11, 77, 192
   tectonic, 12

velocity discontinuity, 194

Whataroa River, 204
windowing, 124

ζ-calibration, 50